2nd Edition

User Interface Evaluation & Design

James N. Helfrich, Ph.D.

Rylan

Published by Rylan Books
Ronkonkoma, NY 11779

ISBN 10: 1-60797-896-2

ISBN 13: 978-1-60797-896-1

Printed in the United States of America.

This book is printed on acid-free paper.

Print Number 5 4 3 2 1

Table of Contents

Unit 1: Overview of Usability 4

01 History 5
02 Interaction Framework 14
03 Components of Usability 25

Unit 2: Analysis 41

04 Target Audience 42
05 Task Analysis 56
06 Criteria Identification 66
07 Heuristic Analysis 78

Unit 3: Design 88

08 Executive Summary 89
09 Design Process 97
10 Prototypes 108
11 Specification 121

Unit 4: Usability Studies 136

12 Experimental Evaluation 137
13 Data Collection 151
14 Qualitative Analysis 162

Unit 5: Standards 169

15 Layout 170
16 Color 186
17 Typography 201
18 Web Applications 219
19 Mobile Applications 231
20 Desktop Applications 244

Appendix 261

Glossary 262
Works Cited 269
Index 271

Overview of Usability

This unit describes the fundamental principles of HCI on which we will base the rest of the class. Though these principles are small in number, each plays a critical role in understanding usability challenges around us.

01 History

The story of Human Computer Interaction (HCI) began before there were computers as we think of them today. The field of study that preceded HCI was "human factors" or how people interact with machines. While HCI has evolved a great deal since its human factors roots, many of the underlying principles originated there.

Usability as a Science

Figure 01.1 - Coal worker shovel

The humble beginnings of HCI can be traced to Pennsylvania steel plants at the end of the 19th century. At that time, manual labor accounted for most factory work; automation was still in its infancy. Workers would physically shovel coal into massive blast furnaces to keep temperatures hot enough to make steel. Initially, workers were asked to bring and use their own shovels. Each worker considered his shovel choice personal; the managers only cared that enough coal was shoveled into the furnace to make the steel. This all changed when Fredric W. Taylor entered the scene.

Scientific Management

The process of increasing work efficiency through analysis of workflow

Taylor is credited as being the father of **scientific management**, the process of increasing work efficiency by carefully analyzing the process workers go through to perform their labors. Taylor theorized that workers could would be more efficient if they were given optimal tools for the job. To test this theory, he conducted a series of experiments with different shovel size, shape, and handle length. Taylor theorized that these variables would influence how long it takes to move a unit of coal and how much effort it takes for the workman to perform the task. These experiments showed Taylor the optimal shovel for moving coal, quite different than one optimized for shoveling sand or snow. While the workers were initially quite reluctant to give up their personal shovel, they found that productivity was tripled using one Taylor provided (Taylor, 1903).

Before Taylor, usability engineers worked with "rules of thumb" and "best practices" where choices between design alternatives were left to personal opinion or individual taste. Today usability engineers approach their craft in much the same way scientists approach theirs: using the scientific method. Theoretical frameworks are carefully created to explain what is observed and these frameworks are then tested with experimental data. This technique allows usability engineers to make far more accurate assessments of usability challenges and be far more confident about their recommendations.

Time & Motion Study

Figure 01.2 - Time & Motion

Scientific management took another step forward when Frank and Lillian Gilbreth refined Taylor's technique to find the best way to lay bricks. As with coal shoveling, brick laying techniques were considered highly personal to the layman (a layman is literally one who lays bricks. Due to the Gilbreths' work, layman has come to mean "unskilled worker"). Frank and Lillian did a time and motion study of the process of laying bricks and revealed that, on average, a layman performed 18 individual motions to place a single brick. They then went about designing the layman's workspace to minimize the number of motions required to lay a brick by carefully placing materials, altering the design of tools, and simplifying motions. When this workspace and technique was used with the same laymen as before, they were able to lay a brick in 4.5 motions (from the original 18) and increased productivity from 150 bricks an hour to 350 (Gilbreth & Gilbreth, 1917).

These same techniques can be applied to the computer realm. Users can be made more productive by reducing unnecessary steps and simplifying the rest. This requires usability engineers to do the same thing Frank and Lillian did for the laymen: carefully observe the user's workflow and design a workspace to improve the layman's performance.

Task Saturation

Task Saturation

The condition when the user does not have the time to perform the necessary tasks even when he knows how

When Orville and Wilber Wright set about building the first flying machine more than a hundred years ago, their primary concern was the design of the airplane itself. The pilot was an afterthought. In fact, the first design had the pilot lying on the top of the wing, craning his head back for a view of where the plane was headed. As airplanes become more powerful and complex, the needs of the pilot could no longer be ignored. During World War II, accident investigators began to notice that most airplane accidents were caused by pilot error, not broken airplanes or weather. Even though pilots knew what actions needed to be performed in a given situation, they simply did not have time to perform those actions quickly enough to save the airplane. This is called **task saturation**.

To address this issue, designers had to turn the human-factors question on its head. Instead of modifying the human to fit the needs of the machine, the machine was changed to fit the need of the human. Two designers, Fitts and Jones, began to look at cockpit design with the same scientific approach Taylor used to optimize shovel design. They found that by simplifying certain controls, making important status messages more prominent, and reducing distractions, airplanes became safer (Fitts & Jones, 1947) (Fitts, 1954).

Fitts' Law

The closer the control and the larger the control the easier it will be for the user to select it

As part of this research, Fitts and Jones realized that the difficulty of reaching a given control is a function of the distance a pilot needs to reach and the size of the control itself. A small control close to the pilot's hand is equally easy to reach as a large control further away. This relationship is called Fitts' law. It turns out that Fitts' law is equally applicable to the computer realm as it is to the cockpit; small buttons close to the user's cursor are equally easy to click as large buttons on the other side of the screen. **Fitts' law** is represented with the following equation:

$$\mathrm{ID} = \mathrm{Log}_2(1 + \frac{\mathrm{D}}{\mathrm{W}})$$

Here, ID is the index of difficulty or how hard it is to click a button. D is the distance the cursor must travel and W is the size of the button to be clicked.

Ergonomics

Fifty years ago, there was no such thing as an "information worker." This is because most people were employed to work with their hands to some

Ergonomics

The process of adapting the work environment to fit the needs of the human frame

capacity, be that in a factory or on a farm. After the Second World War, more and more people were employed to generate, manipulate, or consume information. About this time, secretaries started complaining about wrist pain due to hours spent typing on a mechanical typewriter. Doctors started noticing loss of eyesight due to hours squinting at x-rays with poor lighting conditions. Businessmen started experiencing back and neck pain due to sitting behind a desk all day. These ailments were the result of people adapting to an unfamiliar work environment: the office. Applying human factors principles gleaned from Taylor and others, scientists learned that the amount of physical effort required to perform a typical information worker task could be greatly reduced with small adjustments in posture, lighting conditions, and desk design. These and other influences are collectively called **ergonomics**.

Ergonomics plays a large role in how we interact with computers. A primary example of this is eyestrain. For many years, it was assumed that people who worked with a computer monitor were doomed to wearing glasses. The reason for this chronic nearsightedness was due in part to the constant distance between the individual's eyes and the screen they viewed for hours. If we spend too much time working at the same focal distance, we lose the ability to focus at different distances. This particular problem can be addressed by the user periodically focusing on things far away, by not placing the computer screen against the wall, and by carefully monitoring computer time. Another source of eyestrain is a result of the computer screen itself. If the contrast between the background and the screen is too large (either due to a too-bright monitor or a too-bright background) then eyestrain will likely result. Finally, if a reflection appears in the computer screen, the eye will subconsciously adjust focus from the screen to the reflection. This rapid refocusing can damage our focusing ability. Therefore, care should be taken to place a monitor in a setting with moderate light, facing out a door, and with no reflections.

Another example of ergonomics in the office is typing efficiency. Keyboards were initially arranged alphabetically in 1867. The first widely adopted keyboard design was the QWERTY (named for the sequence of characters along one row) layout developed by Sholes in 1874, the design chosen to avoid jams associated with other layouts. In 1932, Dr. August Dvorak performed a series of studies revealing that the QWERTY keyboard caused us to perform most of our typing with our weakest hand (56%), makes most use of our weakest finger, and most of the common sequences of letters in the English language require us to jump over the home row. The result of this study is the Dvorak simplified keyboard, designed specifically to reduce fatigue, errors, and typing time. Unfortunately, his design never caught on because the improvements were not deemed to be worth the changeover cost (Dvorak, 1936).

A final example of the principles of ergonomics applied to computer work is the position of the computer screen with respect to the user's eyes. When the screen is too high or low, the neck muscles are strained and worker fatigue is drastically increased. Similarly, keyboards placed higher than the user's elbows result in arm strain and wrist injury. This problem has been largely addressed with modern office furniture design but has unexpectedly resurfaced with the introduction of mobile devices. Because the best viewing position is necessarily higher than the best hand position, there is no good way to interact with a phone.

Futz Factor

With the surplus of computing power available on the typical personal computer, designers began adding customization and personalization features to software in an effort to make computers more fun. While it was hoped these features would increase productivity, the opposite proved to be the case. In a 1996 study by the Gartner Group, it was found the typical information worker spends 5.1 hours a week playing with their computer. What makes this study remarkable is that it was conducted long before there was widespread access to the Internet in the workplace. This unproductive work is called "Futzing."

The futz factor is a form of escapism where the user unintentionally dwells on unimportant details rather than tackle difficult or unpleasant tasks. Typical examples include spending a dozen minutes adjusting the font on a resume, repositioning windows on the desktop, and gratuitous scrolling on a mobile device. Each of these has little impact on the task at hand yet offers a welcome diversion to the user. Today, designers attempt to make using software a rewarding and fun experience without offering too many unneeded distractions in an effort to minimize the futz factor.

Human-Computer Interaction

Strictly speaking, the field of HCI lies within that of human factors. What makes HCI unique, however, is that cognitive factors dominate HCI problems. Advances made in our understanding of HCI parallel the migration of the computer from an expensive research curiosity to an integral part of our everyday life. Indeed, many HCI problems can be said to be rooted in this migration. Early computer systems were developed by highly educated and technical people to be used by highly educated and technical people. Predictably, when these same systems were foisted on the general population, problems resulted.

The first set of problems came from the adaptation of paradigms and metaphors from the programming world. Variables, hierarchies, loops, and pointers are concepts familiar to all programmers. To most users, however, these paradigms are foreign and a little intimidating. Perhaps it should not be surprising that usability problems result when programmers choose to use these paradigms to address user interface problems such as file systems and word processors.

The second set of problems resulted from the increasing capability of the computing devices themselves. In order to justify new product purchases and upgrades, software engineers have felt pressured to add functionality and more advanced features into every product. The result has been "feature creep" and "bloat-ware," making even confident computer users feel intimidated and lost from time to time.

The final set of problems was the result of the miraculous penetration of computing capability to the general public. While it took almost a century for electricity and phones to be commonplace, it took a decade for computers to reach the same penetration. As a result, an extremely diverse set of people find themselves using computers every day. Adapting enormously powerful computers running tremendously sophisticated software to nearly every person on the planet is extremely challenging. Airplanes are much simpler to operate than a personal computer yet require years of training and formal certification processes before a would-be pilot is allowed to fly. Yet somehow we expect our children to be able to use our computer. This is our challenge.

The HCI Process

Human-Computer Interaction, from the perspective of a software engineer, is about three fundamental processes: analysis, design, and verification.

Analysis

The process of discovering the problems with a given design

Analysis is the process of learning about the problem to be solved and discovering the issues with a given design. This involves understanding the principles of human-computer interaction, understanding the user and how the user interacts with the system, and identifying the criteria that must be met for a given design to fulfill the needs of a given set of users. Activities performed during the analysis phase include learning about the target audience, understanding how the target audience uses the system to accomplish a task, and identifying what success for a given design looks like.

Design

The process of channeling creative processes to maximize the chance a good design will be found

Design is the process of channeling creative processes to maximize the chance that a good design will be found. To accomplish this, a set of tools have been developed to help the designer focus on the most important components of the design. These tools include a structured, iterative design process through which multiple design alternatives are evaluated. Following this, one or more prototypes are created where the strengths and weaknesses of a design can be more readily identified. Finally all the design details are carefully described in a functional specification.

Verification

The process of validating all the assumptions made during the analysis and design process

Verification is the process of validating all the assumptions made during the analysis and design processes. This typically involves taking a design to real users and carefully observing how they interact with the system. The steps in the verification process include identifying the research questions that need to be answered in the usability study, creating experiments that enable us to gather data for these questions, combining all the experiments into a single usability plan, performing the plan on members of the target audience, and carefully analyzing the results.

This book is structured around these three processes. As each of these processes is studied in detail, we will apply what we learned to create a product specifically for a given set of users we identify. The goal here is not only to master all the tools and techniques of HCI engineering, but also to increase our confidence that we can use them in the workplace.

Examples

The purpose of these examples and problems is to help us better identify usability challenges in the world around us. We will begin with examples from the physical world because they tend to be more obvious than software examples. As the semester progresses and our analysis tools sharpen, we will move to more subtle problems.

Example 1

Light Switches

Figure 01.3 - Light Switch

Problem:
Imagine a classroom with two banks of lights. The first bank consists of most of the lights in the room while the second is the single light over the door. The purpose of two banks is that the instructor can dim the lights for the purpose of viewing the computer screen without making the room completely dark. For the convenience of the instructor, there are two sets of switches: the first is by the door and the second is behind the podium. What is right and wrong about this design?

Solution:
One thing right about this design is that an individual entering the room can control the lights and the instructor at the podium can control the lights. In other words, both of the locations where control needs to be exerted have a copy of the light switches.

The first problem with the design is that "up" does not necessarily mean "on" for these switches. If the door switch is set to "up" then "up" means "off" for the podium switch. The opposite is true of the door switch is set to "down." Thus, the "up means on" metaphor is broken for this design. It is replaced with a less-common "change switch means change lights."

The second problem is the placement of the two switch controls relative to each other for a single switch set. At the podium computer, the control nearest the door operates the lights above the door and the control farthest from the door operates the rest of the lights. For the switch set nearest the door, it is less clear what each individual control does. People typically figure this out by experimentation. To make matters worse, some of the classrooms have the controls swapped so, even if you do learn the placement in a given room, you will get it wrong in the next.

Example 2 **Cell Phone Locking**

Figure 01.4 - User on phone

Problem:

Rachel's phone has a locking mechanism to prevent pocket dialing. The phone has another feature which unlocks the phone when a text or a phone call arrives. This was done for convenience reasons: why make the user open the phone when a text arrives? What is right and wrong about this design?

Solution:

One thing right about this design is that it is easy to lock and unlock the phone. In fact, when a text arrives and the phone is on a table, it takes no effort to read the text.

Unfortunately, if the phone is in Rachel's pocket, then an incoming call or text will unlock the phone. This makes it highly likely that Rachel will hang up on the caller while she fumbles through her pocket. Worse yet, if a text arrives and Rachel decides to ignore it, then an unlocked phone is in her pocket. This results in many pocket dials because the "talk" button is easy to hit in the pocket.

Comprehension check

Question 1 **Terms & Definitions**

Define each of the following terms:

- Human Factors
- Scientific Management
- Ergonomics
- Human-Computer Interaction

Question 2 **Pioneers**

Briefly explain the major contribution of each of the following human factors pioneers:

- Dr. August Dvorak
- Fitts and Jones
- Frank & Lillian Gilbreth
- Fredric W. Taylor

Question 3 **HCI Activities**

List the steps to solving a typical HCI problem.

Problems

Problem 1

Figure 01.5 - Bank of glass doors

Bank of glass doors

Consider the common bank of glass doors in a public building. A student is walking to the library talking with a group of friends. She walks up to a bank of glass doors with a push-bar. She leans against it while turning back to see what her friend is saying. The door does not budge. That is odd; the library is obviously open because she saw people inside. What is right and wrong with this design?

Problem 2

Figure 01.6 - Door handle that turns

Door handle

The designers of the library doors from the previous problem decided to remove the standard push-bar on the door and use a new design. What is right and wrong with this design?

Problem 3

Figure 01.7 - Scissors

Scissors

Consider the following pair of scissors. The user in this scenario is a teacher needing to cut out many patterns from a stack of construction paper for his first grade class. What is right and wrong with this design?

Problem 4

Keypads

Figure 01.8 - Phone and calculator keypads

James is a Junior in High School and is taking a Physics class requiring a lot of calculation. He tries to touch-type with is calculator but for some reason it won't work. Why? He has been able to touch-type on his phone for years.

Consider the following keypads. The left is for a phone and the right is for a calculator. What is right and wrong with this design?

Problem 5

Electric razor

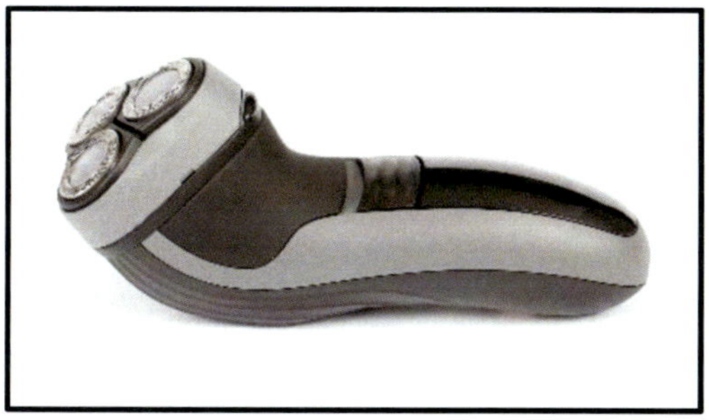
Figure 01.9 - Electric Razor

Tim works at Home Depot and, unfortunately, has the morning shift. This means that he has to be at the store an hour before it opens at 7:00am. Wanting to squeeze every last minute of sleep out of the morning, Tim eats his breakfast pop-tart and shaves in the car. This means that Tim drives to work half-awake shaving with his new electronic razor. What is right and wrong with this design?

Problem 6

High school desk

Figure 01.10 - High school desk

Ethan is a junior at Madison High School hoping to go to college someday. Today is a typical hectic day for Ethan: he has soccer practice after school, needs to find his friends Steve and Ben to plan their weekend, and find Sarah to ask her for a date. He might even find time to do some homework if nothing else comes up. Right now, he is just trying to get to English class before Mrs. Wilson marks him tardy again. As he slides into his chair with 10 seconds to spare, he shoves his notebooks, books, and papers in the basket below his chair. What is right and wrong with this design?

02 Interaction Framework

While it is easy to point out when an interface has failed the user, it is far more difficult to pinpoint exactly why. The more precisely we can diagnose where the failure happened, then the more accurately we can prescribe a fix. To help with this process, we use a structured approach to the analysis problem. We look at exactly how the user expresses his or her intent to the system, and how the interface communicates the system state back to the user. This structured approach is called the interaction framework.

Communication

The first thing to understand is that Human Computer Interaction is essentially a communication problem. While technology is often used as a tool to address problems, the problem itself is a communication one. In a traditional communication scenario, two humans try to share ideas using a shared language. Success in this scenario is largely a function of the mastery of the two participants over their common language.

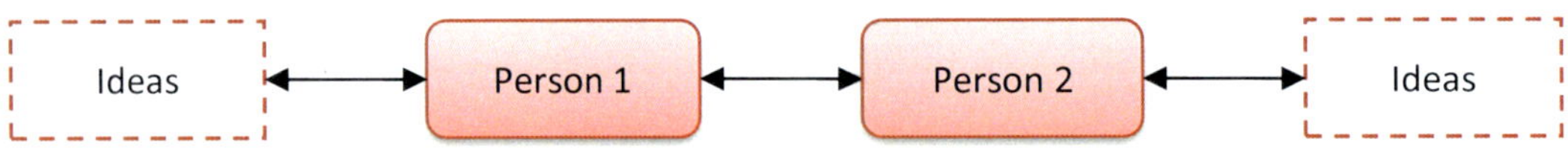

Figure 02.1 - Traditional communication between human participants

HCI, on the other hand, is more complex. Here the interface stands as a proxy for one of the human participants. The designer needs to fully anticipate all the communication needs of the user and to fully express all the communication needs of the system.

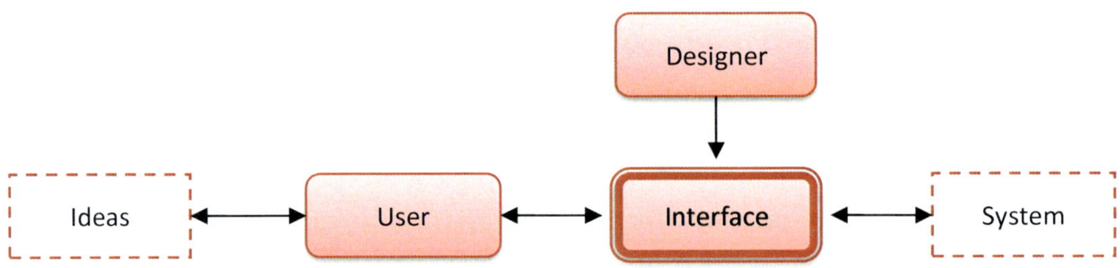

Figure 02.2 - HCI communication where the interface stands as a proxy for the designer

For this to work, the designer needs several things: a deep understanding of the needs of the user, a complete and consistent understanding of the system to be represented, and knowledge of the different available communication options in which the interface will be expressed. If any of these are missing, the interface will likely be a failure.

Interaction Framework

In the simplest terms, the interaction framework is the summation of the input and output channels between the user and the interface. Note that there are two channels: output from the user is the same as input to the interface, and input for the user is the same as output from the interface. Since we typically look at the problem from the user's perspective, we will label these two channels evaluation and execution.

Evaluation is the process of the user interpreting the output from the interface. This channel is completely determined by the capacity of the interface to generate output and the capacity of the user to interpret what the interface has sent. **Execution** is the process of the user expressing his or her needs to the input channels afforded by the interface. Again, it is a function of the capacity of the user to articulate these needs through the input options of the interface.

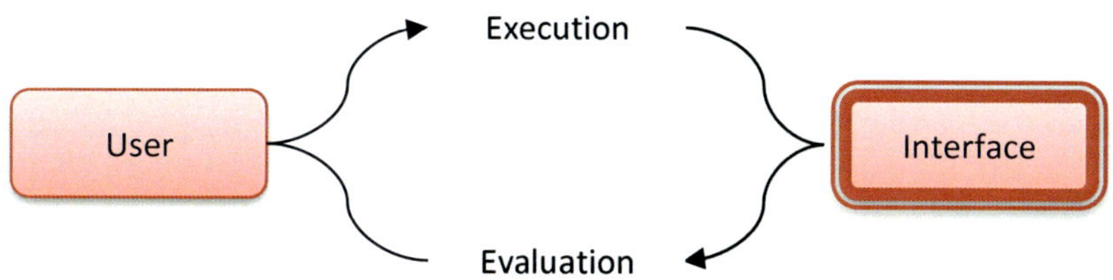

Figure 02.3 - Simple interaction framework. (Norman D. , 1999)

When the interaction framework works, the user is able to fully express his or her needs through the execution channel. Similarly, all output from the interface through the evaluation channel is what the user expects. When one or more of these communication channels breaks down, we have an interface failure. These are called the gulf of evaluation and gulf of execution according to the channel where the breakdown occurs.

Gulf of Evaluation

The gulf of evaluation is defined as the failure of the user to accurately perceive or interpret the state of the system. There are several phases to the evaluation process (Heim, 2007), each of which could be the source of the problem:

Gulf of Evaluation

The failure of the user to accurately perceive or interpret the state of the system

1. **Presentation**: The interface must be able to accurately present the system state and user data.
2. **Perception**: The user must possess the physical ability to accept the input that the interface is presenting.
3. **Interpretation**: The user must have the ability to understand the components of the interface.
4. **Evaluation**: The user must have the ability to relate the interpretations received from the interface to his or her understanding of the system and to his or her goals.

If we were analyzing oral communication, then presentation would be the process of the speaker uttering sound. Perception would be the process of the listener physically hearing the sound. Interpretation would be the user breaking the sound into words, each of which is individually recognized. The process of forming those words into a coherent thought that is understood by the listener is evaluation.

Presentation

The system outputting system state or user data.

Presentation

The presentation phase of the evaluation process is completely up to the system. This is about the system correctly reflecting the user's data or the system state through some interface elements. Often programmers focus on presentation and forget the other elements of the evaluation framework. This becomes apparent when the programmer makes comments like: "The system warned the user, it is not my fault that the user did not know what to do with the information."

Most often, presentation problems are the result of bugs or failure of the system to provide the user with the needed information. To avoid this class of problems, interface designers and software engineers can be tempted to present the user with all available information in the format in which the system contains it. While this may address presentation problems, it often simply moves the problem down to another level of the evaluation framework.

Perception

The user accepting system output through sensory organs

Perception

The perception phase of evaluation is completely up to the user. This is about the user having the physical ability to accept the output that the interface is presenting. This has nothing to do with mental processes; it is a sensory operation. It is somewhat rare to have a perception breakdown be the source of an evaluation gulf.

One example of a perception breakdown is the result of a colorblind user. It is not uncommon for a designer to rely on color to reflect system state. A red traffic light, for example, differs from green only in color. If the user is colorblind, he may not have the physical capacity to perceive the difference. The same is true with deaf users when messages are presented as sounds.

Fortunately, perception problems are easy to detect and easy to fix: present the information to a group of users and asking them if they can perceive the information. If they cannot, make the information more prominent until it crosses the threshold to where all can perceive it.

Interpretation

The user understanding the vocabulary, symbols, and metaphors used by the system

Interpretation

Interpretation is the process of the user understanding the vocabulary, symbols, and metaphors used by the system. In the physical world, interpretation is recognizing that knobs are for turning, handles are for turning, and buttons are for pushing. In the computational world, it is recognizing scroll bars, hyperlinks, edit controls, drop-down menus, and icons. These components are the building-blocks of interfaces, each of which needs to be individually understood by the user.

The most common breakdown in interpretation is when the designer uses a metaphor, symbol, or term with which the user is unfamiliar. For example, a "ding" sound is commonly used to indicate an error has occurred. If the designer relies on the "ding" to indicate an error, the system correctly plays the sound (presentation) and the user hears the sound (perception), an interpretation breakdown could occur if the user fails to associate the sound with an error.

As with perception errors, interpretation errors are easy to detect. Show a group of users the various metaphors, symbols, and terms that are to be used in the interface. If the users misidentify these elements, then interpretation errors are likely to follow. Similarly, one can fix these problems by using more mainstream symbols.

Evaluation

The user relating system information so they achieve an understanding of the system

Evaluation

The evaluation phase is the process of the user relating system information to their interaction with the system or with what they are trying to accomplish. This is a much more complex problem than the other phases of evaluation because it relates to the user's mental model of the system state, the user's intention, and the belief in how things should work. These are all very difficult to identify.

Rather than attempting to enumerate all possible sources of evaluation errors, it is easier to say that if a gulf of evaluation cannot be attributed to presentation, perception, or interpretation, it is probably evaluation. Similarly, addressing evaluation errors is highly situation specific. Usually it involves clarifying the interface so there are fewer opportunities for the user to misconstrue what is going on.

Gulf of Execution

The gulf of execution is defined as the inability of the user to do what he wishes to do with the system. As with the gulf of evaluation, there are several phases to the execution process (Heim, 2007) which must be accomplished for the gulf to be avoided:

Gulf of Execution

The inability of the user to do what he wishes to do with the system

1. **Goal**: Execution starts with the user's goal. Why is he working with the system and what does he want to accomplish? Goals are expressed in terms of what the user is doing, not in terms of the interface or the system.
2. **Intention**: The intention is the user's goal expressed in the context of the system. Intentions are expressed in terms of what the system needs to do to achieve the user's goal. Note that a given goal can be expressed through many different intentions depending on the flexibility level of the interface.
3. **Specify Sequence**: The sequence is the plan of the particular interface actions that the user is to execute in order to carry out his or her intention. The sequence is specific to the finer points of the interface.
4. **Execute Sequence**: The final step is execution of the sequence plan. This involves physically manipulating the various interface elements.

If we were analyzing oral communication, then the goal would be what the user is trying to do. Intention is the process of using verbal commands to a specific listener to accomplish this. Note that written commands, another listener, or even a machine could be used to accomplish the user's goal. Specify sequence is the detailed plan the user will make to accomplish the intention. In this case, it will be the actual words that are to be used. Execute sequence is the physical act of carrying out the specify sequence, in this case it is the physical act of speaking to a given listener.

Goal

What the user wishes to accomplish

Goal

The goal step in the execution sequence is what, in a broad sense, the user is trying to accomplish. This is completely up to the user and independent of the system. In most cases, gulf of execution breakdowns are not due to goal problems because we generally assume that the user is always right. However, occasionally the user is trying to do something which is impossible. In cases like these, no system can accommodate his needs. The best we can do is to inform the user what is possible and what is not.

Intention

How the user incorporates the system to accomplish the goal

Intention

The intention step occurs when the user incorporates the system to accomplish his goal. This is the first step when the system is taken into account, though at a high level. For example, the user's goal may be to get to work in the morning. This seems reasonable. If his intention is to use a mobile card game to accomplish this goal, we have an intention failure.

Intention failures are caused when the user attempts to use a tool in a way in which it was not intended. This may be because the user does not understand the tool or that the tool advertises functionality it does not possess. In either case, no action carried out by the user will accomplish his goal with the intention.

There is no sure way to identify or address all possible intention breakdowns because a great deal depends on the user's perceptions and beliefs. These breakdowns can be minimized, however, if the interface accurately advertises available functionality. In other words, if a design suggests that it can do more than it can, an Intention failure is a likely outcome.

Specify Sequence

The plan the user creates to accomplish the intention

Specify Sequence

Specify sequence is the process of the user formulating a plan to carry out his or her intention with the interface. Note that this plan does not have to be fully flushed out at any one point in time; constructing the plan and executing the plan may happen simultaneously. However, when a breakdown occurs because the user attempts to do the wrong thing, then a specify sequence breakdown occurred.

Many gulf of execution breakdowns can be attributed to a specify sequence failure. Some are the result of the user having an expectation of the presence of controls or the behavior of controls that does not reflect what the system provides. Others are the result of the inability of the user to identify the multitude of intermediate steps required to accomplish his or her intention. Some breakdowns even occur when the user fails to appreciate the myriad of different options or information required to perform the desired task. The common denominator is that the user's plan is insufficient to complete the task or is flawed in some way.

Fortunately, the designer has many tools at his disposal to address specify sequence errors. If the task the user is trying to accomplish (the intention) is well understood, the designer can arrange the controls required to accomplish the task in a linear or obvious way. While the user may still make mistakes, the design can minimize the frequency and severity of these errors.

Execute Sequence

The user physically interacting with the system

Execute Sequence

The execute sequence stage occurs when the user knows what needs to be done and attempts to physically perform the task. These breakdowns are common in the physical world. The basketball player knows he needs to put the ball in the basket but does not have the accuracy to make the shot. The runner knows he needs to complete the race in a certain time to win but does not possess the strength to achieve it. The figure skater knows how to perform a triple axel jump but cannot land it often enough to win a competition. In each case, the steps to perform the task are known (specify sequence) but the execution is too difficult.

Execute errors commonly have two sources. The first is that the precision demanded of the user is unreasonably high. This could be a one or two pixel hit region for a mouse, or a 10 pixel hit region for a finger. Often precision errors are the result of timing requirements: multiple operations need to be performed within an unreasonable amount of time. The second source occurs when the user has less dexterity than the average user. This may be the result of inexperience, patience, physical disabilities, or environmental circumstances. In each case, presenting a design to members of the target audience and asking them to perform an operation will provide insight as to whether execute challenges may exist in a design.

In most situations, fixing execute sequence problems are easy. Making a design more forgiving, expanding hit regions, and broadening timing requirements can typically be employed to address most issues. In each case, the capabilities of the members of the target audience as well as the situation in which they are likely to interact with the interface need to be fully considered.

Parting Thoughts

In the case of an interface failure, it is important to remember that regardless of the exact cause of the breakdown, it is ultimately the interface's fault and not the user's. Perhaps this is best explained with an analogy. Imagine a restaurant owner making food to his taste. "It is not my fault my potential customers don't have good taste" he says as he smugly goes out of business. In the end, the customer is always right. The designer must take full responsibility for all aspects of the interaction framework and do all that he can to ensure communication happens and the user's needs are met. In other words, it is not OK for the designer to say "The user should have read the manual" or "It is not my fault that the user did not notice or read the error message." At the end of the day, if the user is not satisfied with the quality of the design, he might just find another design that better fits his needs.

Examples

Example 1

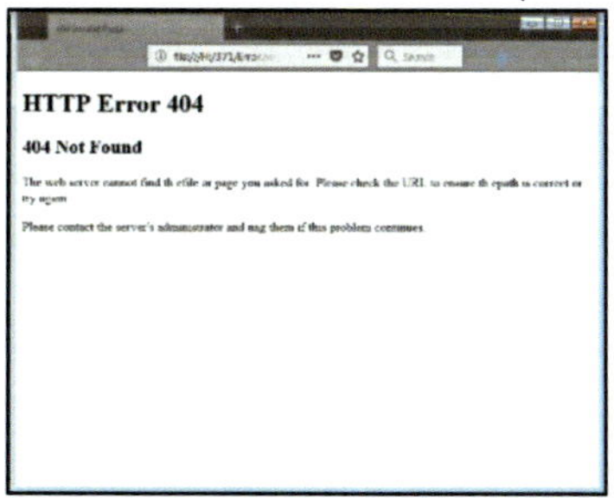

Figure 02.4- Browser with error

Gulf of Evaluation

Consider a user attempting to access a web page. Unfortunately, the user typed the wrong address and the system needs to present an error message. This will be analyzed according to the four evaluation steps of the interaction framework:

- **Presentation** involves the interface creating an error message and presenting it to the user. Since the underlying problem (a typo) is unknown to the system, the system can only respond with what it does know (web page not found). This is presented to the user in the form of a traditional 404 error message.
- **Perception** involves the user noticing the error message. Unless the browser window is occluded by another window, the presence of an error message (and absence of the desired web page) will be fairly obvious to the user, so perception is not the problem.
- **Interpretation**, involves the user understanding that this is an error message. This requires the user to actually read the message, a step skipped by many users. Assuming the user can read and takes the time to do so, interpretation is not the problem.
- **Evaluation**, involves the user relating what she understood from the error message to what she was trying to do. In this case, the user will need to figure out the cause of the error message. Could it be a bad network connection? Could the web site be down? Possibly the browser itself is malfunctioning. Since the user failed to check the spelling of the web address, she failed to correctly evaluate what the system was trying to tell her. This is the source of the gulf of evaluation.

Example 2

Figure 02.5 - user on phone

Gulf of Execution

Consider a user attempting to find a given song in a huge playlist on her phone.

- The first step is the **goal**: to play a song. Goals are expressed completely in terms of the user's need, so this could be achieved with the radio, a streaming service, a CD, or through the music app on her phone.
- From here, the user specifies her **intention**, namely to use the downloaded song on her phone. Since her phone has the song loaded on it, intention is not the source of the gulf of execution.
- The user then **specifies** the sequence and **executes** it in an interactive way. She begins by looking for the song by title. As she scrolls through the list, she realizes that she does not know the exact name of the song. Then she searches by artist. Again, she does not find the song. Finally, she remembers that she put the song in her "Workout" playlist. After scrolling through the entire list, she finds it near the bottom. Thus, we have a failure of specify sequence yielded a gulf of execution. Of course a different sequence might have solved the problem (remembering the exact name of the song) or even a different intention (using her streaming service instead). This means that the problem could be addressed at more than one level.

Comprehension check

Quiz 1 **Seatbelt**

Identify the source of the breakdown the following scenario: I get into my car and start driving down the road. After a minute or two, I notice a chiming noise coming from my car. This noise is unfamiliar and I am unsure where it is coming from. I nervously ignore it and continue to my destination where it disappears.

Quiz 2 **Table Cell**

Identify the source of the breakdown the following scenario: I am trying to make a table cell in my word processing document larger. As I move the mouse to the corner of the table, the icon briefly appears indicating I can drag the cell. Unfortunately, as I attempt to click, my mouse moves over one pixel and the icon disappears. This happens several times until, in frustration, I give up.

Quiz 3 **Library**

Identify the source of the breakdown the following scenario: Jack wants to watch the classic film "Gone With the Wind" so he walks to the public library. While the library has a few movies available, it does not nearly have a collection as complete as the books and magazines. Of course, that movie is not present.

Quiz 4 **Doorbell**

Identify the source of the breakdown the following scenario" While baking cookies I realize I am a half cup short of the necessary amount of flour. I go to my neighbor's house to "borrow" some. I ring the doorbell but do not hear the familiar chime I expect. Instead, I hear soft music coming from inside the house. I ring it again and again, still not hearing the familiar chime. Finally, I start knocking. After a few moments, my neighbor answers the door and asks me what is so urgent: he heard me ring the bell three times and knock.

Quiz 5 **Icy Roads**

I am driving my car up a steep hill in the winter. For some reason, the car is slowing down even though I want to accelerate. I press the accelerator harder but the car continues to slow. I jam it down all the way to the floorboard, hearing the engine race faster and faster. Still I slow to a stop; the more pressure on the gas pedal should make my car go faster!

Problems

For each problem, list the eight steps of the interaction framework. Make a case why there is or is not a break in each step for each problem.

Problem 1

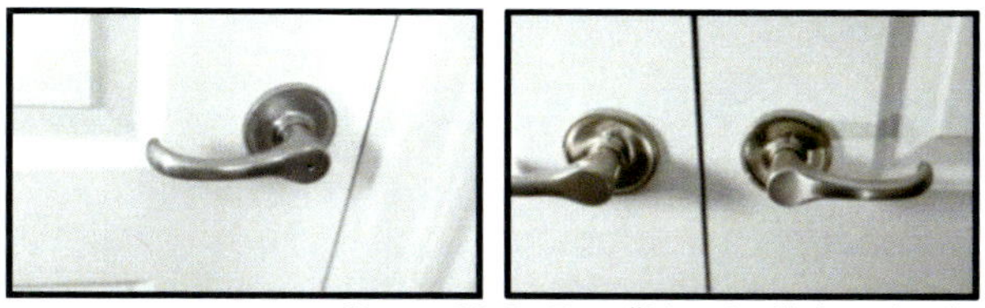

Figure 02.6 - Two "identical" door knobs

Door knob

There are two styles of door knobs in my house. The first is for doors that latch, such as bedrooms and bathrooms. These door handles require the user to turn the knob to open the door. The second is for closets and they "snap" into place, not having a latching mechanism like the others. These handles do not require a turn of the knob, but rather for the user to tug at the door. What part of the interaction design framework is the source of the problem?

Problem 2

Figure 02.7 - Water fountain

Drinking fountain

Timmy is two years old and attempting to get a drink of water from the drinking fountain. He has seen adults, like is dad, do this many times. They lean against the fountain and water comes out! Now is his turn. He walks up to the fountain and leans against it. Sure enough, water comes out! As he gets on his tip-toes to reach the water, however, the water stops. Confused, he takes a step back and tries it again. Same result. This goes on for a couple minutes until he gets too frustrated and asks his father for help. What part of the interaction framework is the source of the problem?

Problem 3

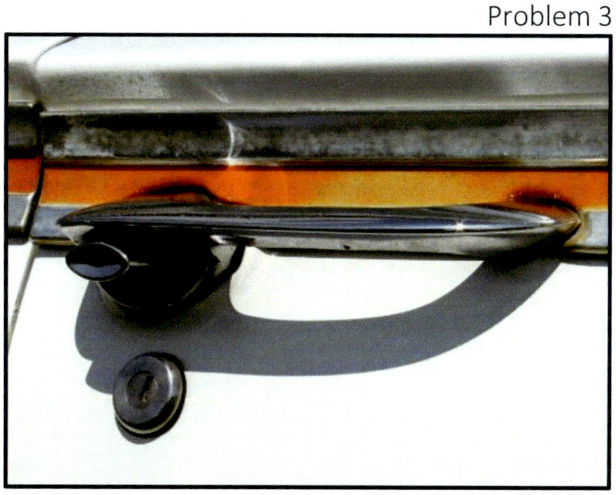

Figure 02.8 - Car door handle

Car door handle

Jimmy is five years old and likes it when his father calls him a "big boy." One day, Jimmy's father asks him if he would like to ride along to the store. "Sure thing!" he exclaims and runs up to his old truck in the driveway. As his father climbs into the driver's seat, Jimmy tries to open the door. At first, he grabs it with his right hand like his dad. That doesn't work! Then he grabs it with two hands. After a few moments, he gives up and waits for his father to reach over and swing open the door. So much for being a "big boy." What part of the interaction framework is the source of the problem?

Problem 4

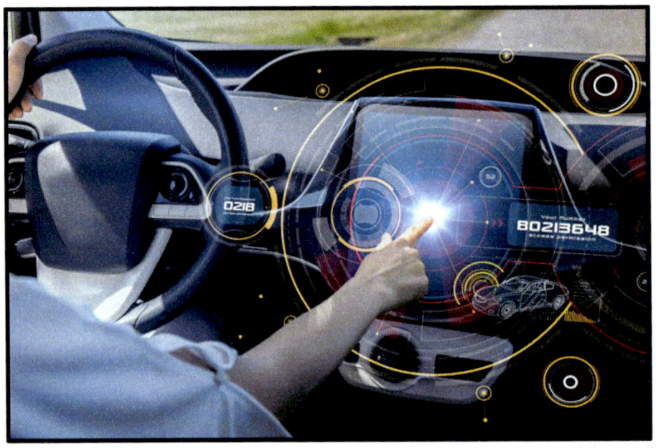

Figure 02.9 - Dashboard

Dashboard

It is a busy day on the Los Angeles freeway and Tony is exhausted. After such a tough day in the office, why does he have to put up with 90 minutes of traffic in the I-5 as well! All he wants to do is be home with his family and relax. Tony is driving his one-year-old car that has a glass cockpit. All the controls are on a large touch screen and all the car's status information is displayed there as well. As he battles the stop-and-go traffic, he periodically skips songs on his playlist and switches back to the map to see if there is an alternate route. This requires Tony to take his eyes off the road and often he makes a mistake in the selection. Woah! He almost got in an accident when he was looking at the screen. "I have got to be more careful!" Tony thought to himself as he refocuses on the road.

What part of the interaction framework is the source of the problem?

Problem 5

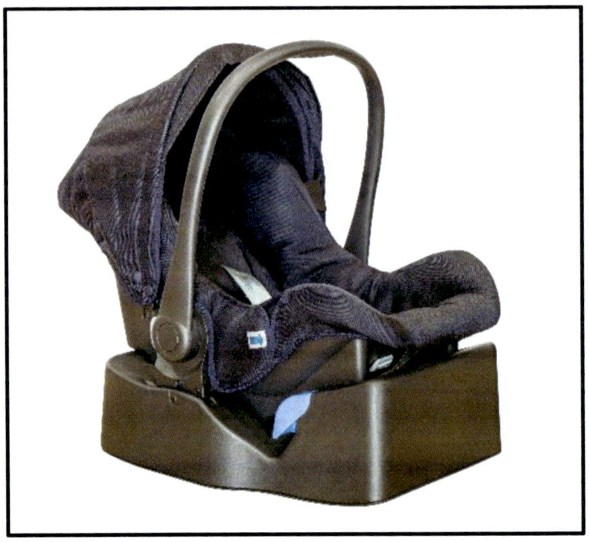

Figure 02.10 - Car seat base

Car seat base

A young mother pushes a full grocery cart out into the parking lot. An ice cold rain is coming down in sheets so she tries to hurry. In the front of the cart is an infant car seat with a sleeping 3-month-old inside.
As she approaches her car, her baby wakes up crying: she wants her bottle. Meanwhile a car has followed the mother because she wants her spot. The paper bags are getting wet.

The mother fishes through her purse for the car keys, hurriedly opens the door, and tries to get everything the car as soon as possible. As she plops the car set in its holder, it doesn't latch. She tries again, but it doesn't latch. The baby is screaming, the groceries are getting wet, and the waiting car is getting impatient.

What part of the interaction framework is the source of the problem?

03 Components of Usability

There are many ways to measure "quality" for a given software product. Some researchers have identified dozens of components or variables in the quality equation. Examples include: resistance to change, reliability, maintainability, relevance, and so on. When trying to ascertain the quality of a given component or system, it is useful to look at each of these variables individually. One component frequently listed in quality descriptions is "ease of use" or "usability." Through the years, researchers have identified many sub-components to this. These can be distilled into a list of eight. These eight, called the components of usability, are efficiency, learnability, familiarity, simplicity, mapping, motivation, trust, and visibility (spelling "ELF's M&M TV"):

Efficiency	The amount of effort or time required to perform a task.
Learnability	The path to becoming proficient.
Familiarity	The degree in which the interface resembles something with which the user has previous experience.
Simplicity	The amount the user needs to know to master the system.
Mapping	Clues within the design encouraging the user to form a consistent mental model of the system.
Motivation	How does the system make the user feel?
Trust	The amount of confidence the user has when using the system.
Visibility	The degree in which the functionality and the data of the system is available to the user when he needs it.

Components of Usability:

A structured approach to analyze and design a user interface

The components of usability offer a structured approach to analyzing and designing a user interface. It breaks the problem down into eight axis with which one can measure the successes and challenges, enabling one to compare options and analyze individual elements. In other words, the components are not a substitute for intuition, common sense, and domain-specific knowledge, but rather a framework in which to analyze and approach it.

Which component is most important? Well, that depends on the application, the user, and the scenario. A keyboard is by far the most efficient way to enter data into a computer for an experienced typist. However, when exploring a web page, that same user may prefer to use the mouse because it is more fun. Simplicity might be the key characteristic for a news web site, but most will rank trust as the most important for a bank's site.

Efficiency

Efficiency:

The amount of time or effort required to perform a task

Efficiency is a measure of how much effort is required to perform a task. Possibly Jef Raskin put it best:

> *A computer shall not waste your time or require you to do more work than is strictly necessary.(Raskin, 2000, p. 6)*

Efficiency influences usability because it determines the maximum amount of work the user can accomplish. The process of inputting text into a computer is one example where efficiency concerns are paramount. Most people can think faster than they can text, type, or even speak. A novel interface allowing a user to input more text in a given amount of time would be desirable. Efficiency is an important metric for usability because it is directly correlated with the productivity potential of the user. Believe it or not, users have more important things to do with their time than use our software. Generally their goals are "do these three things" rather than "toil over the computer for hours." There are several factors that typically influence efficiency:

Flexibility Studies show that systems that are adaptable to the user's preferences tend to be more efficient than those requiring the user to adapt to meet the needs of the system.

Directness Shortcuts, keystrokes, and similar mechanisms can make experts significantly more efficient than systems lacking these features.

Responsiveness A system that reacts quickly to user input can make an intermediate user out-perform an expert user using a less responsive system.

Though most measurements of efficiency (such as time to completion, work throughput, resources consumed, etc.) lend themselves to absolute measures, efficiency from the HCI perspective is fundamentally a relative measure: it compares the output of a user against some norm or standard. From this observation, the following scale is used:

Score	Description
2	Users of the interface are more productive than with any comparable design.
1	Overall user performance is better than most similar designs.
0	Productivity of members of the target user group is typical when compared to similar designs.
-1	Overall user performance with the interface is below average or some task yields the worst performance possible.
-2	Users of the interface are less productive than with any comparable system.

Learnability

Learnability:

The path for a novice to become an expert

The learnability is the path a novice user takes to becoming an expert. It is not the amount that needs to be learned (simplicity) nor how pleasurable may the journey be (motivation). Instead, learnability is a function of the difficulty of the learning process. Learnability commonly plays a pivotal role in usability because it directly influences the degree of expertise the user has achieved in using the system. If the learning curve is too difficult, it is unlikely the user will ever become more than a novice. A well designed video game is a great example of an easy learning curve. The longer the user plays the game, the more proficient he or she becomes. In many cases, the user doesn't even realize that learning or work is being accomplished. A poorly designed game, on the other hand, sees the user "plateau-ing" in his or her skill. Here, more game play does not yield more proficiency. The only way to get better is to get coached from a friend or another source.

To determine the difficulty of the learning curve, it is necessary to identify the inductive leaps the user is required to make to reach the next level of proficiency. Consider a modern word processor like Microsoft Word. A novice user will quickly discover the relationship between typing and text appearing on the screen. A bit of exploration will yield the functionality of most of the toolbar buttons. Initially, the learning curve is very smooth. Understanding the Save command, however, requires an understanding of files which may or may not be familiar to the user. Styles and Mail Merge are two features requiring huge inductive leaps. Many complex concepts need to be understood before either of the features can be learned. Unfortunately, there does not appear to be a way to teach these concepts to the user from within the application. For this reason, the dividing line between the intermediate user and the advanced user is typically whether he understands Styles and Mail Merge. This part of the learning curve is so difficult that few users are able to cross it on their own.

The goal of managing the learning curve is to help all users achieve proficiency in the most painless method possible. The best examples are the pay-as-you-go variety. They offer the user an increasing amount of reward as her or her proficiency increases. As the user progresses, an increasing amount of the functionality becomes available until the user has achieved proficiency. To achieve this it is necessary to manage every step of the curve including day zero, partial proficiency, and expert:

Day Zero Day zero is the first exposure the user has to the interface. What will that experience be like for him? Will he be able to muddle by however inefficiently? While it is important to design the interface for the expert user, no one was born an expert. Everyone needed to go through that day zero experience. If the zero day experience is bad enough, there will be no day one.

Partial Proficiency What is the user experience like in the middle of the learning process? Is the next step obvious? Is the usability at this step acceptable?

Expert Is there a point when the user has learned all that he can about the product? If so, will continued exploration give him frustration and disappointment or satisfaction?

When evaluating the learnability of a given system, the following scale is to be used:

Score	Description
2	A member of the target user group is almost certain to achieve expert status.
1	It is likely virtually all the users will achieve an intermediate level of proficiency or a majority will achieve expert status.
0	It is likely that a few of the users will achieve expert status or a majority of the users will achieve intermediate proficiency level.
-1	It is probable that a few members of the target user group will achieve an intermediate level of proficiency.
-2	It is unlikely the target user will advance from the stage of novice after exposure to the system.

Familiarity

Familiarity:

The degree in which the interface resembles something the user has used before

Familiarity is a function of how similar a given design is to other designs the user may have encountered before. When familiarity is properly leveraged in a design, a novice user can move directly to an expert status without encountering the learning curve. Familiarity is critically important in applications where a high percentage of users are novices. An example would be an e-commerce web site where the user is not using the site long enough for learning to occur. There are two main components to familiarity: consistency and compatibility.

Consistency Consistency is a measure of how one part of a design resembles another. For example, do all the dialogs in a given application have a [cancel] button or do some have [close] and others have [quit]? By maximizing consistency, it becomes easier for a user to master unseen parts of the application because all the parts work the same. Think of all the different ways there are to open a door: swing out, swing in, swing up, slide to the right, French-door style, automatic doors, and countless different latching mechanisms. Would it drive you crazy if every door in the university was different and before entering a room you had to stop and figure out how this particular door works?

Compatibility Compatibility, on the other hand, is a measure of how one design resembles another. For example, the steering wheel on all cars behaves the same. If one car used a steering wheel, another used a tiller like a ship, and a third arrow buttons like a game controller, then there would exist compatibility problems.

Designing for familiarity requires knowledge of the user's "visual vocabulary." There are some things which all humans can be expected to be familiar with. Most users additionally have a vocabulary specific to a given system. Most Windows users, for example, are familiar with right-click menus and most Macintosh users are familiar with inertial scrolling. Finally, some users are familiar with more advanced metaphors such as block selection and collapsible tree structures. A common mistake made by interface designers is to assume the target user has the same vocabulary as the designer. This, of course, is almost never true because designers have among the most advanced vocabularies.

When analyzing a system for familiarity, the following scale is used:

Score	Description
2	The interface is indistinguishable from other elements with which the target user has experience.
1	The elements of the interface are similar in design and usage to other interfaces the user is likely to have encountered before.
0	The interface is completely novel or shares design elements with interfaces the target user is unlikely to have encountered.
-1	Previous experience is likely to mislead the user or be counterproductive.
-2	Previous experience will severely handicap the user's attempt to use the interface.

Simplicity

Simplicity:

The amount the user needs to know to use the system

Simplicity is a measure of how much the user needs to know to operate the system. A more precise definition is the size and complexity of the minimally consistent mental model. In general, the more the user is required to know, the less usable the system is. It takes years of training and study, for example, to be able to fly a commercial jetliner. A five-year-old, on the other hand, can figure out a go kart in a couple minutes. The primary reason for this difference in usability is that the pilot needs to know so much more to operate the plane.

While the simplicity of the design may be influenced by the inherit complexity of the task, the task does not dictate the simplicity level. For a given design, there may exist many consistent **mental models** for how the system works. It is the job of the designer to pick the most appropriate one. Consider the accelerator for a car. Most users have the mental model "pressing the accelerator makes the car go faster." For the vast majority of all driving situations, this model is sufficient. A driver on the snow and ice may need a slightly more detailed model: "pressing the accelerator makes the wheels spin faster." Drivers of a stick-shift will need to expand their mental model to include the notion of a gear box and will understand that the accelerator only makes the engine spin faster. Finally, a mechanic will realize that the accelerator just pumps more gas into the combustion chamber which, under normal circumstances, yields a faster spinning engine and acceleration to the car. Each of these mental models are consistent to degrees and are appropriate to different classes of users. A common mistake is to assume that the most detailed mental model, or the model the designer first came up with is the most appropriate for the user.

When analyzing the simplicity component of an interface, the following scale is used:

Score	Description
2	The design presents the smallest and simplest mental model currently known.
1	The mental model suggested by the user interface is smaller and simpler than most comparable designs.
0	Members of the target user group are highly likely to be able to internalize the mental model of the system.
-1	Important elements of the design are understandable to most elements of the target user group.
-2	The mental model required of the interface exceeds the capacity of the target user.

Mapping

Mapping, like simplicity, focuses on the user's mental model of the system. While simplicity is a measure of the size and complexity of the mental model, mapping refers to how well the interface communicates the intended model. It is important to note that the user is ultimately responsible for his or her own mental model; the best the interface can do is to encourage the user to form a consistent one and discourage an inconsistent one.

Mapping:

The process of how the interface communicates the system model to the user so he can form a consistent mental model

With any mapping problem, three components must be considered: the desired (and presumably consistent) mental model the designer wishes to project to the user, the current (and likely incomplete) mental model the user possesses, and the way in which the interface design can communicate the desired model to the user. The first component is related to the simplicity discussion in the preceding section, namely that it should be intentional and deliberate on the part on the designer. The second component is much more difficult to understand. Each user is likely to have a different perception of how the system works and is unlikely to be able to accurately describe his understanding to the designer. A skilled usability engineer spends a great deal of time and effort to understand the user and get a better idea of what their mental model looks like. The final component to the mapping equation is how the interface can communicate an accurate mental model to the user. This can be accomplished in three basic ways: through the use of metaphors, direct education, and through deductive reasoning.

Metaphors A metaphor is a bridge communicating a mental model to the user through abstract or representative constructs. Typically a metaphor is a simplified schematic (either static or interactive) of the desired mental model. Consider the B, I, and U buttons on the typical word processor toolbar. The user may believe they mean Backspace, Indent, and Undo. However, by rendering the buttons like Bold, Italic, and Underline text, the user interface is suggesting a different meaning. Metaphor-driven mapping is typically easy for users because learning happens at a subconscious level.

Direct Education Direct education is the process of simply telling the user how the system operates. This could be as simple as labeling controls or as complex as requiring training or certification. Typically direct education is inefficient and requires more effort than other mapping techniques. However, when correctly done, it leaves little room for inconsistent mental models.

Deductive Reasoning Deductive reasoning is the process of giving the user sufficient clues as to how the system works that he can figure it out for himself. Most users do not like solving problems and find that deductive reasoning gets in the way. For this reason, it is generally a bad idea to rely on deductive reasoning for mapping challenges.

When analyzing the influence of mapping on a given usability problem, the following scale is used:

Score	Description
2	It is highly likely that the target user will form a consistent mental model of the system.
1	The interface encourages the user to form a consistent mental model and discourages the user to creating or retaining an inconsistent mental model.
0	The interface does not encourage or discourage any particular mental model.
-1	The interface suggests an invalid mental model or fails to encourage the user to form a consistent one.
-2	The interface discourages the user from forming a valid mental model of the system.

Motivation

Motivation:

The degree in which the user wants to use the system

Motivation is a measure how much the user wants to interact with the software. This includes how rewarding the interface is, how aesthetically appealing it is, and how fun it is. Some of the best usability examples are successes because they are so much fun to play with that the user doesn't notice or care about efficiency. The functionality of a Ferrari is less than that of a minivan assuming that you obey the speed limits. That being said, why is a Ferrari so much more appealing? Is it possible to drive a Ferrari around the block without a huge smile on your face? Of course the emotional appeal of a car is a big factor in a purchase decision. The same is true for your clothes, your house, and your cell phone. But how does that affect usability?

Consider a driving video game. The efficiency is low (it takes a long time to complete a race), the learnability is lengthy (it takes hundreds of games to become an expert), familiarity is poor (each game has a different physics engine and buttons are used to steer the car rather than a steering wheel), and simplicity is lacking (the physics of the car is quite complex). However, because the game is fun and rewarding, I keep playing. I don't even notice the other problems.

Motivation is a **force multiplier**. Consider a tanker used to refuel a jet fighter near a combat area. The tanker itself does not deliver munitions nor are they particularly cheap. However, a tanker enables a fighter to loiter near the combat area much longer than it could do otherwise. In other words, one tanker makes a squadron of fighters have the combat presence of twenty squadrons. It is a force multiplier. Motivation operates the same way in interface design. When the interface is fun, responsive, inspiring, and attractive, the other shortcomings have less weight. In some cases, the failures become less glaring. In others, they are simply forgiven. This is why operating systems, phones, and cars focus so much on aesthetics.

In the past computers were very good at giving the user negative feedback. Every time the user did something wrong, the computer would beep or throw up an error message outlining his shortcomings. This made for a very dysfunctional relationship and somewhat unrewarding experience for the user. A motivating interface, on the other hand, is constantly telling the user "good job." Hardware folks have long known about the importance of ergonomics in driving users' perception of their designs. Your keyboard gives a very satisfying "click" when you press the key. That sound was engineered. The car door gives a satisfying "thud" when it closes. That sound was also engineered. Who thought that typing a letter or closing a door would be fun?

There are many tools that can be used to influence motivation:

Aesthetics Software that is visually appealing tends to invoke more positive emotions than software that is bland or boring.

Rewards Many tasks are intrinsically rewarding. Software can also be used to make mundane tasks seem rewarding. For example, a software interface may allow the user to check off milestones leading to completion of a task.

Gamification Gamification is the process of introducing elements of game play (competition, score, etc.) into non-gaming situations. This can encourage users to complete tasks or use features that they might otherwise not be inclined to use. A key aspect of gamification is to make sure the reward structure is compatible with the mission of the software.

When analyzing the influence of motivation on a user interface, the following scale is used:

Score	Description
2	Members of the target user group enjoy using the interface, yielding feelings of satisfaction, pleasure, and similar positive emotional responses.
1	The user interface is likely to appeal to the target user.
0	The user is neutral about the user interface, exhibiting no emotional response or perceiving a balance between the good and bad components.
-1	The target user would prefer to not use the interface.
-2	The target user's dominant emotional response to the interface is strongly negative, such as: distaste, irritation, insult, or impatience.

Trust

Trust:

How confident the user feels when interacting with the system

Trust is a measure of how confident the user feels while interacting with a system. Poor trust often results in poor motivation and poor efficiency. For example, most users have a high degree of trust in a hand-held calculator. Because nothing could go wrong, because there is no opportunity for data loss or embarrassing consequences of incorrect decisions, most users feel no inhibition to experiment with new features or to use the device casually. Your income tax form falls in the other extreme. Because there is no feedback, because the consequences are severe for errors, and because the terminology is often unclear, users often spend an inordinate amount of effort double-checking their work before submitting results. There are three ways to increase trust in a system: increase system feedback, increase fault tolerance, and help the user feel in control.

Feedback Feedback is the process of keeping the user in the loop with regards to what is going on. This can take several forms, including acknowledgement of an action or status of an operation. In all cases, this is giving the user the ability to monitor what is going on. In other words, the user has the right to demand that the system "return and report." The relationship of a piece of software to the user should not be dissimilar to that of a manager with an employee. While the manager can ask the employee to do certain tasks, ultimately it is the manager who is responsible for ensuring that the task was correctly performed. How does the manager fulfill this responsibility? Generally, he needs to monitor the progress and performance of the employee. A foolish manager would trust a new employee with an important task without providing any means to ensure that it was done correctly. How can we expect users to take that same leap of faith without giving them any ability to monitor the software? Trust needs to be earned. The easiest way to earn the user's trust is to consistently give the user assurance that the task is being well managed.

Fault Tolerance Often the user's feeling of trust is directly tied to how resistant the system is to failure. This contributes to the feeling of robustness and safety. Think of Isaac Asimov's first law of robotics:

> *A robot may not injure a human being or, through inaction, allow a human being to come to harm. (Asimov, 1942)*

The same is true for software. Fault tolerance increases the user's trust because it offers assurance that his assets are secure.

In Control When a user feels in control then his trust is generally increased because he feels that any outcome is the result of his action. This can be manifest in a feeling of mastery, whereas projecting that feeling tends to make users feel frustrated or demoralized. In general, the more predictable the system is, the more the user feels in control.

One of the most common ways to forfeit the feeling of trust is to make the system "too smart." The poster child for this trap is the Microsoft Office AutoCorrect and AutoFormat features introduced in Office 97. The AutoCorrect feature fixes the user's typing mistakes by recognizing a misspelled word and replacing it with a correctly spelled one. If the correction was the right one, then the user is none the wiser; he probably did not even notice the typo. However, if the correction was wrong, then the user has a bad experience because he rightly states that he never entered that text into the document. The auto features in Office may have served the user well 95% of the time, but it made the user lose the feeling of being in control. As a result, many users saw Office as being "too smart for its own good" and were very cautious using it.

When analyzing a system for trust, the following scale is used:

Score	Description
2	Members of the target user group are likely to be convinced that they have complete control over the system.
1	The target user is highly likely to be relaxed and stress-free when working with the system.
0	The system behaves in a predictable way.
-1	The user is hesitant to use the system.
-2	The user is extremely nervous, distrustful, or suspicious of the system.

Visibility

Visibility:

Ensuring all the information is available when the user needs it

Visibility is the process of ensuring that all the information the user needs is visible at the point in time in which he needs it. This holds true for data and controls. There is, however, a complication as Donald Norman points out:

> *Show everything at once, and the result is chaos. Don't show everything, and stuff gets lost.*
>
> *(Norman, 1988, p. 92)*

Imagine a financial advisor asking a customer to choose between investment A, B, or C without disclosing what the differences are. Note that it might that there is a collection of very important differences between these investments, but the advisor is not providing the data. In fact, the papers describing the differences are in a cabinet behind the desk of the secretary in the other room. Now, by some miracle, our customer is able to find this document after digging through all the documentation in the office as the advisor sits idly by watching the search. The customer next needs to pour through the documents and decipher the legal and financial terminology. Finally, our amazingly patient hypothetical customer is able to choose investment option B. "Great," says our suddenly attentive and smiling financial advisor, "Now let's choose our tax option..." Now I realize that no rational customer would sit through this abuse nor would any financial advisor inflict this type of disrespect to a paying customer, but how often does our software make us do this? Is this experience really any different than using emacs?

Typically there are three components to visibility: reachability, discoverability, and precedence.

Reachability Reachability is like driving in a big city. Even if you know where you want to go, the route is not always obvious. In a city like Seattle, there are lakes, hills, and freeways that ensure that no road continues in a straight line for more than a couple miles. Longtime residents pride themselves on being able to get around all the "sticking spots." Software navigation should not be like driving; there are no natural obstacles to avoid. Information should be in an intuitive and obvious location, and getting to that location should be readily apparent. And certainly, at all points in time, the user should have ample clues to their context so they can orient themselves.

Discoverability Discoverability is the measure of the likelihood or effort required for a user to be able to find the information he is looking for. It is one of the few usability principles that can be directly measured. As a result, it tends to be the one most focused upon.

Precedence Precedence is the measure of how prominent a given item is in the user interface. The rule of thumb is that the more important an item is, the greater the precedence. All pixels are not created equal. Some areas of the screen have a much higher real-estate value than others. While it may not be your goal to draw the user's eye, your design may be an unwitting accomplice due to your choice of visual elements.

Visibility is a **zero-sum-game**. The more precedence you give in some data or feature, the more you are taking away from another. In other words, often the best way to increase the visibility of a given feature is to reduce the visibility of another. When analyzing the influence of visibility on a given design, the following scale is used:

Score	Description
2	All the data and functionality required for the user to perform his core scenarios are present in the most convenient and useful format known.
1	All the data and functionality required to perform an action is present in a single location, though more convenient and useful formats exist.
0	The target user is able to locate all the data and functionality required to perform a scenario without undue inconvenience.
-1	Data or functionality required to complete a scenario can be found by the target user after significant searching.
-2	Data or functionality required to perform an action cannot be found by the target user.

Parting Thought

One thing you may have noticed about these variables is that they tend to be interrelated. Sure you can increase the trust, but what does that do to the efficiency? Improvements in visibility tend to be at the expense of aesthetics. Similarly, breakdowns in one variable tend to result in breakdowns in others. When motivation suffers, learnability is similarly adversely influenced because the user's patience is reduced.

Remember, the purpose of the variable is not to replace creativity or intuition, but to channel it. There is never a replacement for grey matter.

Examples

Example 1 **Light Switch**

Figure 03.1 - Light switch

Problem:

There exists a light switch near the front door to a house. There are two controls: one for the porch lights and one for the light in the living room. Both of these lights can be controlled by other switches elsewhere in the house so "up" could mean "on" or "off" for either switch.

While coming home from work one evening, Sally realizes that that porch lights are not on. When she walks into the front door, she flips the nearest switch. Unfortunately this turns off the lights in the living room prompting her daughter to complain. Embarrassed, Sally turns the light back on and then toggles the other light switch. A few minutes later, Stan walks down the hallway and notices the lights are off there. He flips the switch expecting to turn on the hall light. Unfortunately, this switch turns off the living room lights instead of turning on the hall lights. Now his daughter is irritated, thinking that people are just picking on her!

What variable is the source of the usability breakdown?

Solution:

- Efficiency: This is not the problem because the light switch responds to the user input immediately. (2: can't be improved)
- Learnability: This is not the problem because the user has used this light switch many times before and has thus already learned how to use it. (0: average)
- Familiarity: Both Sally and Stan are conditioned to use the nearest switch to turn on the nearest light. Unfortunately, this is not how their house was wired. Thus, experience will be a disservice (-2: familiarity hinders usability)
- Simplicity: There is a very small mental model, but the mental model is typical (0: average)
- Mapping: The light switch furthest from the door operates the lights closest to the door. Thus the mapping is misleading (-1: misleading)
- Motivation: The user has zero emotional response to the light switch (0: average)
- Trust: The light switch always responds to the user control, but selecting the wrong control is embarrassing. This makes the user hesitant to use the system (-1: hesitant)
- Visibility: The switch itself does not reflect state ("up" does not necessarily mean "on") or indicate which light it controls (-2: important state not presented)

Example 2

Figure 03.2 - Cell phone with user

Cell Phone Locking

Problem:

Rachel is talking with some friends when the cell phone in her purse chimes with the incoming-text sound. Not wanting to be rude to her friends, she ignores the phone. A few minutes later, Rachel hears a voice coming from her purse. She retrieves the phone from her purse to discover that she just hung up on her mother. Not only did she pocket-dial her, but she hung up on her when she retrieved the phone. Why does her phone unlock when there is an incoming text? Analyze this scenario according to the variables of usability.

Solution:

- Efficiency: The phone unlocks easily. (2: can't be improved)
- Learnability: It is not difficult for Rachel to figure out how to unlock the phone. (0: average)
- Familiarity: The locking and unlocking mechanism is similar to other phones, except for the feature that unlocks the phone when a text arrives. (-2: familiarity hinders usability)
- Simplicity: There is a very small mental model, but the mental model is typical (0: average)
- Mapping: Rachel formed an accurate mental model of the locking mechanism based on interacting with the system (1: most users form an accurate model)
- Motivation: The user has zero emotional response to the locking mechanism (0: average)
- Trust: Frequently Rachel pocket-dials after an incoming text or hangs up on a caller while trying to retrieve her phone. This makes her extremely hesitant to use the device (-2: extremely hesitant)
- Visibility: The phone announces its change of locking state with a sound (1: the state change is announced)

Comprehension check

Quiz 1 **Terms & Definitions**

For each of the following definitions, name the corresponding variable of usability

- Clues within the design encouraging the user to form a consistent mental model of the system
- The degree in which the functionality and the data of the system is available to the user when he needs it
- The amount of effort or time required to perform a task
- The amount the user needs to know to master the system
- The path to becoming proficient
- The amount of confidence the user has when using the system
- Does the user want to use the system
- The degree in which the interface resembles something with which the user has used before

Quiz 2 **Variable Details**

Identify which variable is referred to by the following description:

- Common tools to achieve this include Metaphors, Direct Education, and Deductive Reasoning
- A force multiplier
- The time or effort required to achieve proficiency
- A <computer> may not injure <your data> or, through inaction, allow <your data> to come to harm
- Consistency and Compatibility are the two main components
- The Office 97 AutoCorrect feature was widely criticized because it compromised what usability principle?
- A computer shall not waste your time or require you to do more work than is strictly necessary
- Show everything at once, and the result is chaos. Don't show everything, and stuff gets lost

Problems

Problem 1

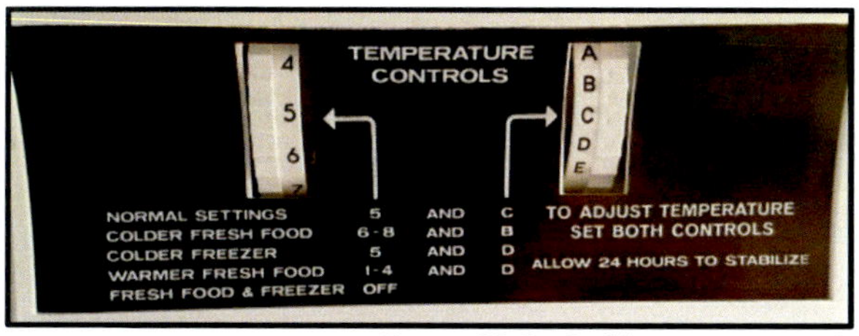

Figure 03.3 - Refrigerator controls

Refrigerator

Jenna opens the refrigerator door to get some milk out for breakfast and notices that there is frost on a couple left-over containers. She then opens the freezer to find that the ice cubes in her tray are melting. Clearly she needs to adjust the temperature controls. In the back of the top shelf, she finds what she is looking for:

Now she is more confused than ever. What should she do to fix the temperature in both the freezer and the refrigerator? Analyze the refrigerator controls according to the variables of usability?

Problem 2

Figure 03.4- Range top controls

Range top

Hannah is with some friends at a recipe party. Though Hannah knows the homeowner Mindy very well, she has never been in her kitchen before. As the party mingles in the kitchen, a pot cooking pasta begins to overflow. Mindy asks Hannah to deal with the little emergency as she is the closest to the range. Mindy can narrow it down to two knobs, but is not sure which one operates the back burner. Meanwhile, more boiling water spills onto the range. Analyze the range top controls according to the variables of usability.

Problem 3

Figure 03.5 - Dvorak keyboard

Dvorak keyboard

Quinn is a Computer Science college Freshman, having used a keyboard for most of his life. In fact, Quinn learned to touch-type in elementary school at the instance of his father. Yesterday, Quinn was working on a project with another student Daniel and they were using Daniel's laptop. Ben offered to type a few of their ideas but, to his surprise, the characters he typed where not the characters that appeared on the screen. Daniel then explained that he configured his laptop to use the Dvorak keyboard because it was much faster than the standard QWERTY keyboard Quinn was using. This intrigued Quinn. As soon as they were finished with their project, Quinn rushed home to learn the Dvorak keyboard. This proved more difficult than he expected. Finally, after several hours of frustration, Quinn gave up. Analyze the Dvorak keyboard according to the variables of usability.

Problem 4

Figure 03.6 - Cursor keys

Cursor keys

Ken is a secretary in a car dealership spends a lot of time on the computer. In fact, it seems that Ken spends most of his days filling out forms. This requires him to use the cursor keys more than just about any other key on the keyboard. One day he came across a new keyboard that has the cursor keys arranged in a plus sign rather than the typical inverted T? Eager to make his job easier, he bought one of these keyboards. After a couple hours, however, he switched back to his old keyboard. Score (-2 to 2) each of the 8 variables and provide a brief rational why your score is correct.

Problem 5

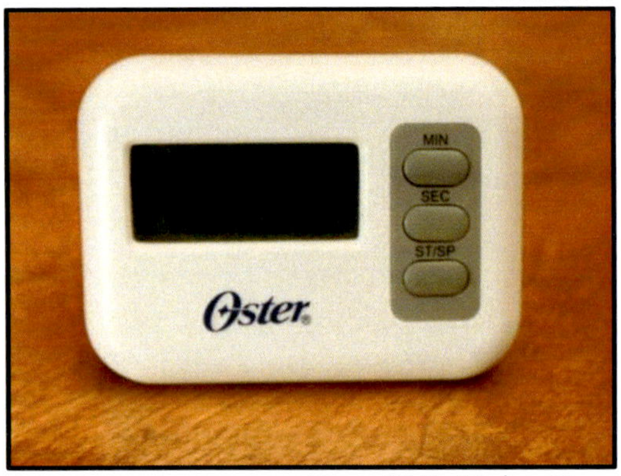

Figure 03.7 - Kitchen Timer

Kitchen timer

James likes to help is wife in the kitchen whenever he can. As he comes home from work one day, he notices that there is more activity than usual: his wife is both cooking dinner and baking a birthday cake. Eager to help, James asks what he can do. His wife instructs him to set the timer for 12 minutes and drop a pound of pasta in a boiling pot of water. He grabs the timer and, seeing it is set for 14 minutes, proceeds to add minutes until it crosses 60 and thereby resets to zero. “Why didn’t you just hit the reset button?” his wife asks. “Reset button?” Score (-2 to 2) each of the 8 variables and provide a brief rational why your score is correct.

Problem 6

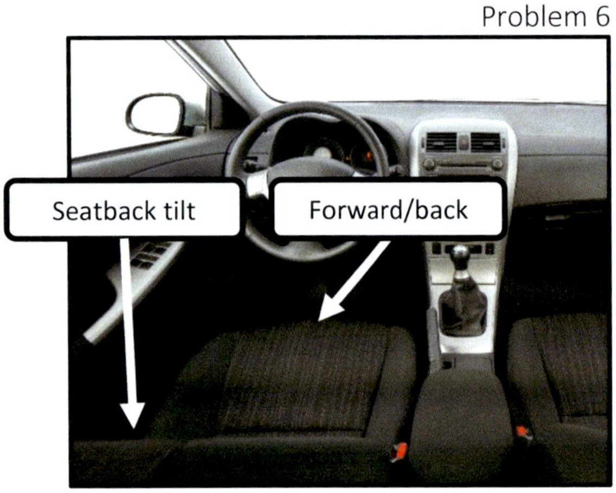

Figure 03.8 - Car seat controls

Car seat controls

Darrell is new to driving and a bit nervous when behind the wheel. This nervousness is doubly so when his father is riding beside him. While they are driving, Darrell realizes that the seat is a little too far forward; it should be moved back about an inch or so. As they pull up to a stop sign, Darrell decides to adjust the seat. He reaches under the seat for the release mechanism and suddenly the seat shoots forward and the car comes to a screeching halt. Darrell looks sheepishly over to his dad who is busy trying to stifle back his laughter. Score (-2 to 2) each of the 8 variables and provide a brief rational why your score is correct.

Analysis

Analysis is the process of systematically looking at a given design in an effort to discover usability challenges.

04 Target Audience

Recall that HCI is essentially a communication problem, finding better ways to allow the interface to explain the system state to the user and finding ways to enable the user to express his or her intention to the system. To successfully design a user interface, it is important that the designer understand both the system and the user. Misunderstanding of either will almost certainly result in an interface breakdown.

It may seem that understanding the user side of the equation is simple: find her and talk with her. One slight problem with this approach: what if all the users are not the same? Worse yet, what if the set of potential users is extremely diverse? It is easy to give up in the face of the complexity of this problem and ignore the user side of the equation entirely. This, of course, is a recipe for a user experience breakdown.

To illustrate this point, imagine a design team working on a product to produce marketing material for a small business. The developer asks "Shouldn't we have customization for the print dialog?" The manager, weary of the costs associated with such a feature, replies "Customization is not necessary for version one." "But someone might want to customize the print dialog," replies the developer. Now the manager is cornered. There is almost certainly going to be at least one user who might want to customize the print dialog. However, in his gut, the manager knows that not many users will want it to justify the cost. What should he do? The problem with the developer's argument is that he was referring to "someone." Who is this someone? Could it be a real user that the developer met? Could it be the developer himself, thinking he is a typical user? Worse yet, could "someone" be something the developer made up to win the argument?

The only way to resolve arguments like these are to agree ahead of time for whom the interface is being built. Not only must the design team agree on the identity of the group of users as well as their characteristics, they also need to be able to internalize this user. The manager calls upon the designer to help with this.

The designer begins by interviewing dozens of small business owners. While they are all unique and have their own challenges, they share a few important common characteristics. All of these characteristics are then folded into a fictional character called Frankie that personifies the typical small business owner. At the next team meeting, the designer introduces everyone to Frankie. First, a brief character sketch is presented. Next, he relates how the team's product will fit into Frankie's daily routine. Immediately the team can see the value of designing the product for Frankie: the target users are basically the same after all.

Later that day, the manager and the developer resume their discussion: "Shouldn't we have customization for the print dialog?" The manager replies, "But customization does not fit into Frankie's workflow and he seems unlikely to want to do such a thing." "That is true," says the developer, "but he will probably want to save as a PDF." "Good point; let's put that on the list."

Observe how the conversation is much more productive and less contentious when everyone on the team is on the same page and trying to design for the same user.

Gathering Data

Everything we do in the design, development, and verification of user interfaces must be based on hard data. It is not enough to think we understand the users! We must constantly gather data to broaden and sharpen our understanding of them.

When talking about "the user," it is easy to get confused about whom we are actually talking about. To clarify that, consider the following definitions:

Customer An individual who makes the purchase decision for a product.

User An individual who uses a product.

Stakeholder An individual with a vested interest in a design. This could be a user, a customer, or even the people building a product.

Target audience The group of people for whom a design is intended. In other words, the collection of all the users.

The first step here is to obtain an unambiguous specification of who the target audience is for a given design. In most cases, this is given to us by the stakeholders (perhaps your manager or the executive sponsoring the project) or the customers (if you are hired to build a product for a specific purpose). There are many times when this is difficult. We need to perform market research to see who is using the product or, worse yet, we need to anticipate who might be using the product once it is finally built. Regardless of the circumstances, the entire design team needs to be on the same page as to who they are building a product for.

The next step is to learn about this group of people. Often we can use research gathered for other purposes to help us here. Marketing may have collected data on the target audience. There may be academic or governmental statistics that is relevant and useful. In most cases, this prior research is not enough. You are going to need to leave your office and interact with real users. You are going to have to enter their workplace and watch them work (with permission, of course). You are going to have to set up formal or informal interviews. You are going to have to seek out experts who have experience working with your users. This is not a small task that can be completed in an afternoon. It is part of a long process that takes several weeks and months.

Once you have gathered all this great data, your job is not done yet. The problem with "the user" is that they are a moving target! The original Apple iPhone was designed for tech savvy early adopters. Just a few years later, they were mainstream. During this short period of time, the technology was being used by senior citizens and junior high school students. If the Apple engineers kept with their original understanding of the target audience, they would not be producing designs that serve the vast majority of their actual users. Thus the prudent designer will devote a few hours every week collecting and analyzing data about their target audience.

The final step in this ongoing and iterative process is to record and share this data. To do this, we need a tool which captures a large amount of diverse qualitative data in an easy to understand format. That format is called a persona.

Definition of a Persona

Recall that the labels "customer," "user," and "stakeholder" refer to real people interested in our products and "target audience" refers to the group of real users. A persona, on the other hand, does not refer to a real user but rather is an abstraction of a segment of the target audience. A persona is a hypothetical archetype representing either the entire target audience or a significant portion of the target audience.

Persona

A hypothetical archetype representing either the entire target audience or a significant portion of the target audience.

Thus a persona is a representation of a group of users. This means several things. First, a persona is based on the user, not the customer or any other stake holder. Second, a persona must share the essential characteristics of the group of people it is meant to represent. This means it must be based on real data. Finally, a persona does not have to be a description of a real person. In fact, a persona is rarely based on a single real person. Most members of a target audience are atypical in some way. The persona, on the other hand, must be representative of the entire audience. For this reason, personas are commonly inspired from one or two actual users. However, they are fictitious because they are aggregates of many individuals. (Cooper, 2004) (Head, 2003)

In the cases when the target audience is a large and diverse set of users, it may be the case that a single persona cannot cover all the bases. In cases such as these, more than one persona may be necessary. With each new persona that is created, the risk increases that members of the development team will not be able to internalize all of them. In other words, the value of the persona will decrease.

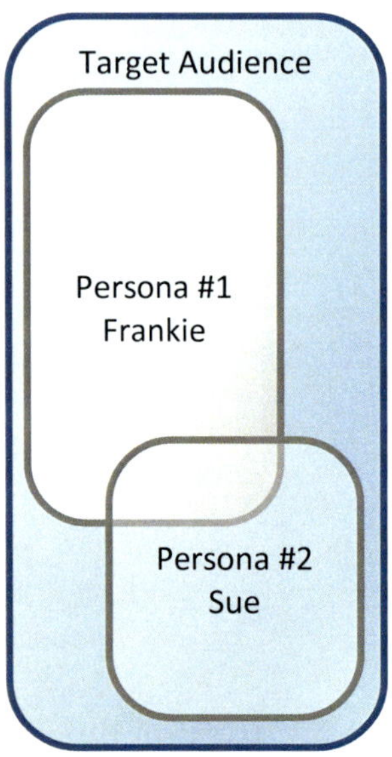

Figure 04.1 - Venn diagram of target audience

Our goal is that the vast majority of the members of the target audience will be covered by the personas we have built. Consider the figure on the left. The blue area represents all the users in the target audience. The first persona, Frankie, covers about half the area. This means that Frankie can stand in for half the target audience. The second persona is Sue which covers a much smaller area. Between the two of them, the majority of the target audience is covered. Notice how there is a small overlap between Frankie and Sue. Generally we want to avoid this. The worst case would be two personas who cover the same members of the target audience. In this case, we have redundancy that could be avoided. Notice also how there are some members of the target audience which is not covered by either persona. What do we do? We could add additional personas, each covering a successively smaller area. We could modify one or both of our existing personas so we get better coverage. We could just be OK with the fact that some fringe members of the target audience are not to be foremost in our thoughts as we analyze and design the interface.

Building a Persona

Figure 04.2 - Notepad and computer

A common misconception is that building a persona is a one-time event. This is incorrect because a persona is never "done." Instead, a persona represents our best understanding of our target audience at this moment in time. In other words, a persona is always a work in progress.

The persona building process consists of the following steps: interview members of the target audience, analyze interview data to find patterns or similarities between interviewees, use these patterns to refine existing personas or create new personas, and search out more data to continue the cycle anew.

Gather data

The heart to the persona building process is data collection. The reason for this is simple: the value of a persona is completely defined by how closely it reflects the audience it is meant to represent. Therefore, if the data is inaccurate or non-existent, then the resulting persona has no value.

Though we will learn more about interview techniques and qualitative data analysis later in the semester, take comfort in the following truism: the only way to not learn something from interviewing a user is by failing to conduct the interview in the first place. In other words, while we will learn how to get higher quality data and how to analyze the data with great sophistication later, you will learn something about your target audience by just observing them.

Organize the data

Once a dozen or two interviews have been conducted, common characteristics begin to emerge. These characteristics are called variables. It is easy to think of variables in the mathematical sense or in the programming sense. These variables are a bit different. Variables in this sense mean attributes, influences, traits, or characteristics that may or may not be quantifiable. Attitude towards computers can be construed as a variable. Web browsing habits may be construed as a variable. Anything that sets one individual apart or that binds groups together can be construed as variables. As we observe users for the purpose of understanding them, we are on the lookout for variables.

Figure 04.3 - Taking notes

Perhaps this is best explained by example. After the first potential user is interviewed, a list describing her characteristics is built. Next, another user is interviewed and a list is created describing his characteristics. At this point the interviewer notices that there are a few attributes from the second user that are shared by the first. The first list is then augmented to correct the previous oversight. With every successive interview, similarities and differences between the various interviewees are noted. This process continues until patterns emerge. It should soon be possible to group similar users.

When we are finished with the organizing data phase of the persona building process, we will have a couple sets of users. Each set is characterized by a few distinguishing variables. Typically there will also be a few outliers that resist attempts to make them members of a set. It may be that we will never be able to group those users or it may be that, with more data, sets will emerge.

Rough draft

Each set of users is a potential persona. Start with the common variables, the ones that characterize the set. Note that these variables alone are not sufficient to build a persona. For a persona to be effective, it must seem alive and real. It must be easy to relate to and be memorable. This requires us to invent attributes that are not atypical of real users who were interviewed yet do not necessarily come from an actual user. The trick is to not make these fictitious attributes overshadow the variables the persona is meant to represent.

A persona is a character sketch consisting of a name, age, quotes, vocation, picture, and anything else necessary to make the sketch come alive. Often they are motivated primarily from a single actual interviewee but more often they are composites of two or three interviewees. When creating a draft, it is helpful to ask if the persona is detailed enough to seem alive and real. Can you imagine having a conversation with your persona?

Verification

At all points in time, we must remember that the persona must represent a large part of the target audience. It must be based on real data and not misleading in any way. There are two ways to verify a persona: to double-check it against the data from which it was built, and to validate it against new interviewees who are members of the target audience.

The double-check process starts with the original data from the interview batch. Start with the set of interviews from which the persona was built. Check the enumerated variables against the attributes of the persona. Are all the important attributes met? Is there anything about the persona that is not representative of each interviewee? How can the persona be adjusted so it is more accurate?

The second phase of the verification process is an ongoing process. The designer needs to make a continual investment in his persona to make sure it is as accurate as possible. He needs to be on a constant lookout for members of the target audience and take every available opportunity to meet with them. Try to assign each one to an existing persona. If one does not fit a persona, add it to the list of outliers. If it does fit, make sure the persona is representative. This may require us, with time, to make minor adjustments to our persona to ensure it is as accurate as possible.

Publishing

The final step of the persona-building process is to share it with others. This can be done in a variety of ways, including distributing print versions in team meetings and creating collages that are hung in the office hallways. In other words, if the persona is not frequently used and referred to by the members of the team, it fails to be an effective design tool.

Some user experience engineers shy away from using personas or sharing them with the development team. This is because they can feel fictitious and "made up," not as tangible as real-live users. There are two ways to address this issue.

First, the onus is on the engineer to tie the characteristics of the persona to actual data. Quotes, statistics, and experiences collected from those we interviewed to understand the target audience should be included in the persona. This "hard data" goes a long way towards making the persona feel more real and representative of those who we are building the software for.

Second, it is often helpful to build a persona from an actual user. Use the user's actual name (or a close variation thereof), and actual picture (with permission), and actual quotes. While building a persona for a team I was working on many years ago, one user really stood out as representative of the larger target audience. Fortunately, this user owned a restaurant that I frequented. One day, I invited the entire team to this restaurant for lunch. The owner came out to greet our party and I introduced him. After he left, I went on to describe him as the basis of our persona. I was careful to point out the two small ways which he was different than our typical user, and how those differences were represented in our persona. After that lunch, the team had no trouble visualizing our target audience

Using Personas

Possibly the best way to use a persona is to treat them as real people, members of the design team. Have mock conversations with them during design exercises. Ask them questions when evaluating design alternatives. You know these efforts are successful when you accidentally call a real user by a persona name.

Analysis

Personas can help with analysis activities because they help us understand the user side of the communication interaction. Consider the interaction framework. While looking for a Gulf of Execution or a Gulf of Evaluation, the usability engineer needs to know if the user would be able to complete a given task. If the user is represented in a well-understood persona, then the engineer would ask the following questions:

- Would Frankie be able to **Perceive** the signal sent from the interface?
- Would Frankie correctly **Interpret** what the signal means?
- Would Frankie be able to **Evaluate** the Interpretation and draw correct conclusions?
- What would Frankie's **Goal** be when using the interface?
- How would Frankie translate his **Goal** into **Intentions** at this point?
- What Sequence would Frankie **Specify** to carry out his **Intention**?

- Would Frankie have the dexterity to **Execute** the sequence he previously specified?

Each of these questions would be difficult to answer unless the usability engineer had a clear picture in his mind of who exactly Frankie is and what class of users he represents. The usability engineer can also follow the same process when conducting a heuristic evaluation using the variables of usability. In this case, each variable is viewed by the usability engineer through the eyes of the persona:

- How likely and how easy will it be for Frankie to learn to use this feature? (**Learnability**)
- Are the design elements of this design likely to be **Familiar** to Frankie?
- Does the complexity of the design exceed the attention Frankie can afford or will it be **Simple** enough for him to understand?
- Will the interface encourage Frankie to form a valid Mental Model? (**Mapping**)
- Will Frankie like the interface? (**Motivation**)

As with the interaction framework, these questions can only be answered if the usability engineer has a clear picture of for whom the design is being evaluated.

Design

As with analysis activities, a persona is an important tool for keeping the design team focused on the correct body of users. This is important because it is all too easy to subconsciously design the product for ourselves rather than for the target audience.

Perhaps this is best explained by example. Imagine a 35-year-old software engineer building a sofa-cushion-fort for his young daughter. Left to his own devices, dad will probably build a fort that he thinks is cool. This includes a front and back door, a balcony created with the dining room chairs, and a large room supported by a sheet and rope hanging from the ceiling lamp (he is quite proud of this design feature). When dad finishes with the fort, he shows it to his young daughter. Unfortunately, she hates it! Dad build a fort for himself, not for his daughter. Undaunted, dad tries again. This time, at every turn, he asks his daughter what she wants. He makes a little room under a table for her stuffed animals. When that is done, she checks it out and offers a few suggestions. He then makes a hallway to another little room where she can have tea with her friends. Again, she checks it out before he moves on. When the fort is finally finished, dad is confident that she will like it because she was involved in every aspect of the design.

Personas allow us to bring the users into the design process. We can ask them questions and get feedback. Of course no real users are involved, but if we are able to internalize them and know them well enough to be able to predict how they would respond, it is the same thing. Personas are an indispensable tool for designers to internalize a large and potentially diverse user base.

Example

Dan the designer has been tasked with creating a web site for a local high school mountain bike team. To accomplish this, Dan begins by building a persona.

Step 1: Gather data

Dan starts by learning as much as he can about the target audience. He begins by talking to the stake holders, the people who are sponsoring his work. At first, they give him very general feedback, "anyone who is interested in the team." After some probing, he came up with a partial list:

Students Anyone who rides for the team.

Parents Parents of the students.

Organizers The coaches, team manager, and sponsors.

Dan starts by working with the students. He formulates a questionnaire which will help him understand this segment of the target audience:

> Demographic questions about age, gender, experience, etc.
>
> What do you love most about riding?
>
> Why did you join the team?
>
> When you want to tell others about the last race or practice, what medium do you use and why?
>
> How do you learn about practice and race information?

After interviewing five students, it becomes apparent that this segment of the target audience actually consists of three sub-segments: the athletes themselves, the friends of the athletes, and potential athletes who are thinking of joining the team. The questionnaire is adapted to these sub-segments and the interviews continue. After a dozen athletes, eight friends, and ten potential athletes are interviewed, Dan is beginning to feel confident that he understands this segment of the target audience.

Next Dan starts working on the parents. His initial questionnaire is:

> Demographic questions about tech competency, number of children in the sport, time spent supporting the children, etc.
>
> Please describe what type of coordination you need to do with other parents to support your athlete.
>
> What technology do you use to perform this coordination? Do you feel that this technology works for you or do you wish you had something different?
>
> How does information currently flow from the coaches to you and vice versa? How could this information flow be improved?
>
> How do you like to receive notifications?

Notice how Dan's questions for the parents are much more open-ended than they were for the athletes. This is because Dan tailored his questions to the maturity level of his target audience, only asking questions that they are equipped to answer. After just two interviews, Dan noticed that a big part of the parents' communication efforts resolved around two activities: sharing pictures and coordinating meals with other parents when they travel for a race. Dan adds a few questions to his questionnaire to better cover these topics.

Finally, Dan prepares a questionnaire for the coaches and team manager:

> Please tell me exactly what information you need to share with the parents and the athletes. For each piece of information, please let me know when the communication needs to happen and how you share the information currently.
>
> How often does the plan for practices or races change? How do you currently communicate those changes?
>
> Is there anything about your current communication avenues that work particularly well or work particularly poorly? Please explain in as much detail as possible.

Based on these interviews, the following data was collected:

Athletes Equal representation across all high school ages and grades. The population is 60% male and 40% female. While the athletes are somewhat interested in the schedule of races and practices, they rely on word-of-mouth to get this information rather than from electronic sources. They do have a high interest in getting race results and referring to old race results. Also, they often look at every single picture taken by friends and parents. This is currently a difficult process because everyone posts their pictures at different locations. Every single picture from a given event is carefully viewed.

Friends Friends share the same demographics as the athletes. Their interest in the team is much smaller than the athletes but they enjoy looking at the pictures. Most of the friends who were interviewed would look at one parent's pictures rather than attempt to find all the pictures taken during a given event. They focus on finding the pictures of their friends rather than browsing the entire collection.

Perspective Athletes For every athlete, there is a prospective athlete (prospective athlete population was estimated to be 110% of the size of the athlete population). They are not interested in the results but are heavily interested in the pictures. Here they want to get a feel for what the races are like rather than looking for specific athletes.

Parents The average parent has 1.25 children in the program. They are mostly interested in logistics (time and location of races, cost of the program, and coordination with other parents). They are also looking for a better way to share pictures with other parents.

Coaches Coaches and managers are very concerned about recruiting (making the team interesting to prorspective athletes) and logistics. They want to post information and want to have a site, but do not want to use the site.

Based on this data, Dan feels like he has sufficient understanding of the target audience and their needs to move on to the next step.

Step 2: Organize the data

After shifting through the data, Dan noticed several variables:

- **Pictures**: Sharing pictures is a big part of the team's communication needs. However, the different populations have different needs. The parents and the athletes want to see all the pictures and they focus on who is in them. The friends, prospective athletes, and coaches are more casual viewers, just wanting to get a general feel for how the event went.
- **Information**: Some users (parents and coaches) are only concerned about logistical information, others don't care about it at all (friends and prospective athletes).
- **Time**: Some users can expect to visit the site for a half hour at a time (athletes) while others can be expected to visits closer to 5 minutes or less (parents, prospective athletes, and friends).

Based on this data, we can see that several segments of our target audience have a significant overlap. The prospective athletes and the friends have similar needs and characteristics. The athletes and the parents also have a large overlap but not enough to place them in the same group. Finally, we realize that the coaches are not really users but rather stakeholders. This yields three distinct groups: athletes, friends, and parents.

Step 3: Rough draft

Our first draft will be for our student athlete persona. We will call her Morgan.

Figure 04.4 - Morgan and friends

Morgan is a 17 year-old student a Madison High School. She is not a big fan of organized sports; why watch other people have fun when you could be doing it yourself? Many of Morgan's friends are also on her team but quite a few are not. They don't really understand mountain biking and why she is gone so many weekends in the fall. Morgan wants her friend to know about the latest race, but does not want to re-post all the pictures on her social media. After all, not everyone cares about her race! Morgan is a typical millennial with a high degree of web and computer literacy.

Next, we need to make a draft for our friend persona. We will call her Amber.

Figure 04.5 - Amber and friends

Amber is a friend of Morgan and has even considered joining the team. It looks like they have so much fun, but she is just not sure that the team is for her. She likes riding her bike and even mountain biking, but racing is a totally different level of commitment. When Morgan posts a picture on social media of her latest race, Amber goes straight to the team web site to see how it went. It looks like so much fun!

The last persona is the parent. We will call her Rachel.

Rachel is a 42 year-old mother of 4. Her second oldest is Morgan, the only member of the team in her family. While she is not "tech savvy," she is not ignorant of technology either. Rachel texts with her older children, has a Facebook account, and regularly shops online. In fact, her husband just bought her an iPad and she has reluctantly started bringing it with her on her daily routine. Rachel is constantly trying to coordinate the activities of her various family members. Why can't they all just be interested in the same things?!?!

Figure 04.6 - Rachel

Step 4: Verification

With drafts of three personas, Dan now needs to verify that they are an accurate representation of the target audience. First, he verifies the athlete persona. Dan meets with the remaining members of the team and discovers something. Several athletes are involved in more traditional sports (two in baseball and one in tennis). Furthermore, almost all of the athletes run track and field in the spring. This potentially invalidates the part of the Morgan persona that refers to organized sports. However, there certainly is a feeling of "counter culture" in the team so he decides to leave that persona unchanged.

Next, Dan interviews several more parents. There is a huge discrepancy in technical proficiency. Does this new piece of data challenge the validity of the Rachel persona? After some consideration, Dan decided to not update this component. There are several parents with low technical proficiency and he would rather build the web site to fit everyone's needs rather than risk excluding a few. There is also quite a bit if discrepancy with regards to the parent's commitment to their children's sports program. Some parents are at every race and are active in supporting the team. Other parents do not respond to e-mails or phone calls. However, the more active parents are the ones who will use the site the most so that component of the Rachel persona remains. In fact, Dan decides to elaborate on it a bit. Here is the new draft: The last persona is the parent. We will call her Rachel.

Rachel is a 42 year-old mother of 4. Her second oldest is Morgan, the only member of the team in her family. While she is not "tech savvy," she is not ignorant of technology either. Rachel texts with her older children, has a Facebook account, and regularly shops online. In fact, her husband just bought her an iPad and she has reluctantly started bringing it with her on her daily routine.

Rachel is constantly trying to coordinate the activities of her various family members. This is particularly challenging with her daughter Morgan because her races are out of town every weekend. To lighten her load, Rachel works with two other parents to carpool and share making meals.

Figure 04.7 - Rachel

Finally, Dan would like to verify the friend persona. While it was very easy to find members of the athletes group and the parents group, the friends is much harder. Dan can't just show up at practice and start interviewing! Fortunately, a few of the coaches have athletes on the team and they were able to introduce Dan to some of their friends. After a couple interviews, Dan became confident that a lot of the athlete data carried over to the friend persona, but there was some minor differences. None of these were significant enough to change the Amber persona.

Step 5: Publishing

Dan presents his research and persona to the stakeholders and is invited by his boss to introduce the team to Rachel, Morgan, and Amber.

Comprehension check

Quiz 1 **Terms & Definitions**

Find a name for each of the following definitions

- The group of people for whom the interface is designed
- One individual from of the group of people for whom the interface is designed
- The individual who pays for the product
- An archetype representing part of the population of individuals for whom the interface is designed
- An individual who has a vested interest in the design

Quiz 2 **Motivation for creating Personas**

What is the motivation for creating and using personas in the design and evaluation process?

Quiz 3 **Example Persona**

Research was conducted where ten individuals from the target audience were interviewed. The following data was collected:

- Average age: 30, where 67% were between 25 and 37.
- Goal: Create an audit trail to avoid litigation, dissatisfied customers, and miscommunications with coworkers.
- Occupation: Information worker. Produces and consumes information. This includes lawyers, architects, marketing professionals, and doctors.
- Technology competency: None are trained to work with technology, but technology has become an important component of their workflow in recent years as the world has become more digital.
- Technology attitude: Early adopters who see computers as a way to gain the competitive advantage. 9/10 were early adopters of mobile devices, 8/10 carry both a laptop and a tablet to work every day.

Create a persona based on this data.

Problems

Problem 1

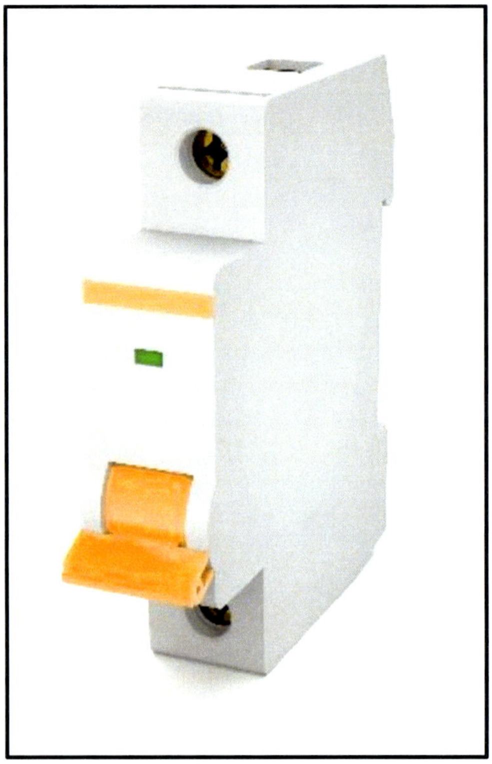

Figure 04.8 - Circuit breaker

Circuit breaker

Identify several personas for a circuit breaker switch:

1) First, create a list of all the possible users for the switch. This list will have about six entries.
2) Next, for each item in the list, give it a one or two word label (such as "Babysitter"). The label should be descriptive, not overly general (such as "female") and not overly specific (such as "14 year old babysitter").
3) Finally, provide a brief rational each persona should be the primary one or why it is less important. Consider who will use the switch the most often, which user will give you the most insight into the usability issues surrounding the switch, and which user will be most dramatically affected by a failure of usability.

Problem 2

Figure 04.9 - Cordless drill

Cordless drill

As with Problem 1, identify several personas for a cordless drill:

1) Create a list of a half dozen users
2) Give each item a label that is descriptive
3) Provide a brief rational why each is more or less important

Problem 3

Figure 04.10 - Stopwatch

Stopwatch app

As with problem 1 & 2, identify the most important persona for a stopwatch app on a mobile device.

1) Create a list of a half dozen users
2) Give each item a label that is descriptive
3) Provide a brief rational why each is more or less important

Problem 4

Figure 04.11 - CB Radio

CB Radio

Imagine you were newly hired to a company that produces CB radios. Knowing your expertise developing user interfaces and feeling that the UI of their current CB radio is a bit dated, they task you with proposing a new design. Your first task, of course, is to create a persona. Please:

1) Describe how you will go about collecting data for you persona
2) What resources will you use?
3) Who will you talk to?
4) How will you make sense of the data you collected?

05 Task Analysis

There are two parts to understanding the user side of the HCI equation: the user and the user's task. While the persona is a great tool for understanding the user, it does not address how the user will interact with the system. This is where the scenario comes in.

To illustrate this point, consider the problem where our task is to understand the usability challenges of a circuit breaker. A circuit breaker is a switch that, with a dozen or so others, resides in an access panel in an obscure corner of the house. When too much power is consumed by a device in the house (say a malfunctioning hair dryer), the circuit breaker cuts the flow of electricity thereby avoiding damage to the house. Our persona for this problem is Marge, a young stay-at-home mom with babies. Marge leaves most home repairs to her husband but, owing to the fact that she is the only adult home for most of the day, finds herself doing some home repair work from time to time.

If the device itself and the persona were our only tools used to understand the HCI interaction, then important details would be missing. We might be tempted to bring Marge into the lab and have her turn the switch on and off to see if she can do it. This would clearly tell us nothing; all people fitting in Marge's persona would probably have no trouble operating the switch. This does not mean all is well with the circuit breaker design, however. We are missing large parts of the interaction framework.

Definition

The scenario is a story representing an important use-case of the user interacting with the system. This represents the second half of the persona/scenario equation, picking up where the persona left off.

Use Case

A key part of the scenario definition is "use-case." A use-case is one instance of a user carrying out his Intention with the system. Recall from the interaction framework, an Intention is how the user plans to use the system to achieve his goal. Often many steps will be required for this Intention to be performed, each of which is represented in the specify sequence part of the framework. Typically, for a given design, there are many use-cases. A sample of the use-cases for the circuit breaker example may include:

- Install the switch
- Turn off a circuit so work can be done on the house
- Turn on a circuit after a problem has been fixed
- Replace a malfunctioning switch

Each one of these use-cases represents what a real user may do with the interface. Note that all parts of the interaction framework are covered here. Take, for example, the "install a switch" use-case. The user will have a goal (finish work for the day and get home to his family), intention (attach the circuit breaker to the breaker box and wire it up), specify sequence (attach the wires by feeding them in through the leads and closing the leads with a screw-driver, then screwing in the breaker to the box), execute sequence (use his screwdriver to do exactly what he planned), presentation (the wire lead interface displays the status of the wire being connected or not), perception (the electrician looks at the wire lead interface), interpretation (the electrician is trained so knows that the wire is well connected), and evaluation (the electrician knows the connection is sound).

Observe how there can be a large number of use-cases even for a simple interface such as that of the circuit breaker. Not all of these use-cases are equally important. Installing a switch may happen only once for example, and it is relatively uncommon to replace a malfunctioning switch. However, the other two use-cases are far more common. The purpose of the scenario is to distill these use-cases down into one or two stories enabling the design team to focus on the most important interactions of the design.

Scenario

The scenario is a story representing an important use-case of the user interacting with the system. Thus though there may be many use-cases for a given interface, the scenario will focus on just one or two. Another important difference is that the use-case focus focuses on one Intention of the user, the scenario attempts to capture the big picture. This includes the user's goals, any environmental factors, and a narrative of what the user is thinking during the interaction.

	Use-Case	Scenario
Purpose	Capture one Intention	Capture the entire interaction framework
Number	Dozens per feature	One or two per feature
Length	One sentence	A paragraph or two
Composition	Nothing more than a label	A linear narrative of the persona interacting with the system
Use	Helps the design team understand all the possible interactions with the system	Helps the design team focus on the most important components of the design

Identifying the Scenario

Given that one or two scenarios need to stand for a large number of use-cases, how do you choose which use-case on which to base the scenario? The answer is: whichever use-case provides the best lens on the problem.

Each use-case potentially draws focus on a different aspect of the design. In other words, each use-case provides a lens through which the design team can get a better perspective of the design problem. There are typically three considerations that might help in selecting a given use-case for the scenario: most common, critical path, and the high-value use-case.

Most common The most-common use-case is the situation that the majority of the users will commonly find themselves. We can get help identifying this use-case through our interviews and market research. Certainly the most-common use-case needs to be carefully considered, especially if it occurs in most or all sessions. For example, virtually all word processor sessions involve opening a file. This use-case makes a strong case for being in the primary scenario.

Critical-path The critical-path use-case is the situation where, if the user fails, the entire product is deemed a failure. Back to the word processor example, not all sessions involve printing. Printing is thus not the most-common use-case. However, printing is a critical-path. Eventually, all high-value documents will get printed. Any word processor unable to print a document is a failure, regardless of any other advantages it may have. Critical-path use-cases make a strong case for being in the primary scenario.

High-value A high-value use-case is a situation instrumental to helping a consumer make a purchase decision of one system over another. Often it is the motivation for switching products or purchasing an upgrade. For example, when the internet hit mainstream in the late 90's, word processors were still used primarily for print documents. When the save-as-html feature was introduced, it was not a most-common use case (few users were using word processors for web pages; most HTML was still created by hand in a text editor) nor was it critical-path (few workflows involved HTML at the time). However, being able to save as HTML was perceived as high-value, elevating this use-case as a strong contender to be the primary scenario.

At the end of the day, the scenario is just a design tool; we never deliver a scenario to the customer as a finished product. Thus it behooves us to ask the following question again and again: how can I make this tool help me to better design the product for the user.

Writing a Scenario

The scenario takes root in one or more use-cases, put together in a single narrative describing a user's interaction with the software. Scenarios have several properties.

- **Linear time**: Time is linear in a scenario. A fictional narrative may incorporate flashbacks and other literary devices. This may be useful in literature to keep the audience's attention, but a scenario has a different purpose: clarity. Thus time is always linear in a scenario
- **Single pass**: The scenario does not attempt to cover all situations or eventualities. Instead, it describes a single instance of the user interacting with the system. When it comes time for a user to make a decision, pick the "representative" course of action and continue. There are no "ifs" in a scenario.
- **Plausible**: The purpose of the scenario is to help us better understand how the user interacts with the system. If the representative interaction is unlikely or not based on the behavior of actual users, the scenario is of little value. Worse! It could be misleading. As with the persona, great effort must be spent to validate the scenario.
- **Memorable**: The persona and the scenario are designed to leverage people's ability to internalize people and stories. To maximize this potential, it is important that the scenario is memorable. An interesting angle, a unique turn-of-phrase, or a well-chosen word goes a long way to making the scenario more effective.

Parts of a Scenario

As with all aspects of HCI, there is no formula for analysis or design. There are often, however, frameworks that help us ask the right question at the right time. The same is true with the scenario. It is common for a scenario to have the following seven components, but it is rare to find a scenario having all seven components clearly delineated. The seven components of a scenario mirror the Interaction Framework:

Setting What is going on while the interaction is taking place? We need to answer such questions as: where will the user be interacting with the system? What kinds of distractions will be present? Are there environmental factors such as noise, lighting, or temperature? Are there social factors such as pressure or expectations? If it is relevant to understanding the user's interaction with the system, we incorporate it. If not, we don't.

Actors Who are the humans in the setting? These include the user as well as others who may play a role. If we are using a persona as an actor, just call them by name; don't redefine.

Goal Motivation of the actors to perform some action. These ties in with the interaction framework. While we don't focus on the goal in the scenario, the fundamental question of "why is the user interacting with the system" needs to be addressed at some level. It could be stated in a sentence or two or implied by the actor's actions. This is a combination of the goal and intention of the interaction framework.

Plan How the actors will carry out their goal. In most cases, this does not need to be stated because it will be implied by their actions. When it is not implied, we need to state it here. The Plan corresponds directly to the specify sequence part of the interaction framework.

Evaluation How the Actors interpret events. Words such as "notice" and "realizes" are commonly used. If the user needs to figure something out that is non-trivial, describe that thought process and what the Actor needs to conclude. Any part of the evaluation side of the interaction framework, from perception to evaluation, can find itself in the scenario.

Actions What the actors do. This is typically expressed with terms such as "clicks", "types," or "opens." Scenarios commonly are dominated by user-initiated Actions. These correspond with the execute sequence part of the interaction framework.

Events External actions or reactions produced by the computer. Some of these are a result of the user actions, others originate from an external source. Usually we need to mention the origin of the event, when it occurred, and what occurred as a result.

Parting thought

If the scenario does not bring insight into how the user interacts with the system, then it is of little value. Ideally, you should be able to say: "if the design helps the user complete this scenario, the design is successful."

Example

Consider the high school mountain bike web site from the previous chapter. There were three personas, but we will focus on one:

Figure 05.1 - Rachel

Rachel is a 42 year-old mother of 4. Her second oldest is Morgan, the only member of the team in her family. While she is not "tech savvy," she is not ignorant of technology either. Rachel texts with her older children, has a Facebook account, and regularly shops online. In fact, her husband just bought her an iPad and she has reluctantly started bringing it with her on her daily routine.

Rachel is constantly trying to coordinate the activities of her various family members. This is particularly challenging with her daughter Morgan because her races are out of town every weekend. To lighten her load, Rachel works with two other parents to carpool and share making meals.

From here, we will list the use-cases; identify which are most common, critical-path, and high-value; identify a scenario from these use-cases; fill in the parts of a scenario; and finally bring it all together into a single narrative.

Use-cases

After reviewing the data from the persona interviews, several use-cases have been identified:

- Discover the date and locations of all the races of the season for the purpose of long-term family planning.
- Learn specifics of the next race the upcoming weekend.
- Find pictures of Morgan from the previous race.
- Determine the cost of Morgan racing next year.
- Connect with other parents to make carpool and meal plans.

Most important use-cases

From the above list of use-cases, the most common, critical-path, and high-value use-cases need to be identified. Based on what we know about Rachel, the most common use-case is "Learn the specifics of the next race." Rachel does this every time she visits the web site. It is worth verifying this assertion with the raw data collected from the persona building interviews. This raw data reveals that parents double-check the race weekend at least once a week.

The critical-path use-case is more difficult to identify. Which use-case did virtually all of the interviewees mention? They all wanted to be assured that they have received the latest info about upcoming races and practices. A breakdown on this use-case is considered catastrophic for everyone who mentioned it.

The high-value use-case is the place where we make a value proposition to the user: why the new design is significantly better than old designs. Most of the interviewees made some reference sharing pictures. If the new web site was able to better facilitate sharing pictures, the parents said they would visit it more often.

Identify the scenario

From these three important use-cases (learn about an upcoming race, receiving notifications, and picture sharing) we need to create a single scenario. Which of these provides the best lens into the usability challenges of the navigation feature? Will one use-case do the job or can/should we fold multiple into the scenario? Is there anything we learned about our persona that will help with this?

The most common use-case (learning about an upcoming race) does involve some navigation but is rather elementary. While it is important, it will not help with the evaluation or design process. The critical path use-case (receiving notifications) also is rather elementary. Finally, the high-value use-case involves some simple navigation but by itself is not very involved. The important thing to realize is that Rachel will probably do all three of these things in the typical session. In other words, our scenario should probably involve several use-cases.

Our draft scenario will have the following components: navigate to the web site, look for any urgent notifications, double-check the race information for this week, and finally review the picture list to see what has been recently added since the last time she checked.

Parts of a scenario

Before the final narrative is written, we need to make sure that all the components are represented.

Setting At home, on a laptop sitting at the dining room table. There is activity in the room as Rachel's children are coming home from school and getting snacks out of the refrigerator. She is trying to answer a question about their plans for the weekend while dealing with a hundred other things.

Actors Rachel, our persona, is the main actor. Additionally there are several children in the room including Morgan.

Goal Rachel wishes to firm up her plans for the weekend.

Plan Rachel plans on navigating to the schedule part of the web site and moving some of the details onto the family calendar. She also plans on contacting another parent to coordinate the carpool.

Evaluation Rachel will view the top-level navigation control to determine where she currently is in the site and determine where she needs to click to go to her destination.

Actions Rachel will tap (click) and swipe (scroll) to get the site to honor her navigation intentions.

Events The app will update the display to reflect Actions Rachel has performed.

Write the scenario

The final step is to author the narrative so the important components of the scenario are preset, but also so the text follows the linear time, single pass, plausible, and memorable criteria for a good scenario. Taking these into account, the result is:

> Rachel is preparing dinner when her children start coming home from school. While eating a muffin, Morgan asks her about the upcoming weekend. She wants to know who is driving and what they will be eating. Rachel walks over to her laptop on the dining room table and clicks on the Mountain Bike Team web site bookmark. Seeing the top-level navigation control, she clicks on the Schedule tab and quickly finds the next race. As she mentions a few details to Morgan, she looks for any changes or notes from the coach. There are none!
>
> As she talks with Morgan about the carpool plan (Rachel is driving this weekend) and the food (pasta again!), Rachel idly navigates to the Picture Exchange tab. She notices that some new pictures were added to the top of the list. As they firm up their meal plan, Rachel periodically calls Morgan over to look through some of the pictures. A few other curious family members peer over their shoulders as they laugh. Perhaps Morgan's brother will ride next year after all!

Comprehension check

Quiz 1 **Terms & Definitions**

For each of the following description, name the corresponding scenario component

- Occurrences initiated by the system or some external source
- Actors' interpretation of events
- Humans in the setting
- Observable behavior by the actors
- What is going on while the interaction is taking place
- Motivation of the actors to perform some action
- How the actors will carry out their goals

Quiz 2 **Scenario definition**

In a sentence or two, define a scenario in layman's terms.

Quiz 3 **Use of a scenario**

When is it appropriate to use a scenario in the design process?

Problems

Problem 1

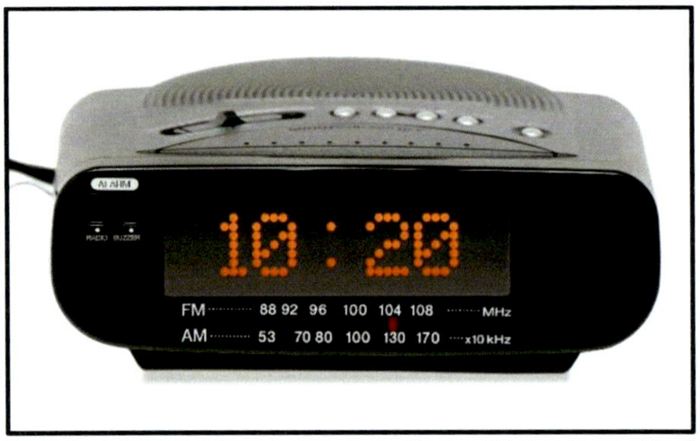

Figure 05.2 – Alarm Clock

Alarm clock

Please identify the most important scenarios for an alarm clock. Identify 3-4 use-cases and provide a short (1 to 2 sentence) description of each.

Persona: Sam is a 24-year accountant who has not outgrown his college night-owl habits. As a result, he never gets enough sleep and always finds himself drowsy while on the job. While he usually does OK at the beginning of the week, he can't keep the yawns at bay on Friday. At least he can get caught-up on Friday! Sam shares an apartment with a bachelor friend from college

Problem 2

Figure 05.3 - CB Radio

CB Radio

Please identify the most important scenarios for a CB radio. Identify 3-4 use-cases and provide a short (1 to 2 sentence) description of each. Use the persona you developed in the previous lesson as your guide

Problem 3

Figure 05.4 - Cordless drill

Cordless drill

Please write a scenario for a cordless drill. Begin by identifying 3-4 use-cases and settle on the single scenario that provides the best lens on the usability of the drill. Finally, write a short (4-5 sentences) scenario. Please use the persona you developed from the previous lesson.

Problem 4

combo atomic 174
complementary 142
contrast 131
coversheet 206
customer 50
deep dive 96
dropdown listbox 174
edit control 175
efficiency 22, 114, 159, 165
elevator pitch 85
enclosing 129
ergonomics 5
evaluation 13, 14
evaluation criteria 71, 82, 93, 97, 115, 116, 118, 119, 124, 126, 189
execute sequence 16
execution 13
executive summary 85, 97, 118, 126

Figure 05.5 - Index from this textbook

Index feature

Please write a scenario for the index feature for an electronic textbook. For this problem, please use the following persona:

Persona: Steve is a 22-year-old college student studying software engineering. He is a diligent student but, due to time constraints, often finds the need to take short-cuts on assignments just to "get 'er done!"

Problem 5

Figure 05.6 - Stopwatch

Stopwatch app

Please write a scenario for a stopwatch app. Please use the persona you developed from the previous lesson.

06 Criteria Identification

How can one tell which of several designs is the best? In this sense, "best" is defined as serving the user's needs. There are many metrics which could be used, but the variables of usability are designed to be particularly useful in this context. They are generic, working with virtually any interface design challenge. There is one problem, however, with using the variables of usability as the sole metric by which one critiques a given design: they do not take into account what we know about the user and the user's task.

The next step in the analysis process is to define success for our given interface in the context of our target audience and the tasks our target audience is trying to accomplish using the system. We call these specialized metrics or heuristics evaluation criteria.

Evaluation criteria are yardsticks allowing us to measure the quality of a given design. They serve to constrain the design process by not only illustrating which design is superior for a given persona and scenario, but also by indicating the direction a design needs to travel to improve.

Criteria v. Requirement v. Feature

The single largest mistake novice designers make is to confuse evaluation criteria for requirements or features. There are two reasons why this is typically the case. First, novice designers commonly start with a design in mind and then try to come up with criteria matching their design. This perspective is limiting and should be avoided for obvious reasons. The second is that it is much easier to think in terms of technology than the user's needs. When we think in terms of the user's interaction with the feature then criterion become much easier to identify.

Feature A feature is a description of one aspect of the project

Requirement A requirement is a description of what must be present in a product

Criterion A criterion is a constraint in the design process through which various design alternatives are measured.

A **feature** is a description of one aspect of the project. A feature is a noun. The end of the design process is typically a set of features intended to meet the user's needs. Criteria are not features. Instead they are the ruler by which we measure whether a given feature is helpful for a user. Features by their very nature are not useful in the evaluation process because they do not provide any insight as to what is good or bad about a given design. It should be noted that a given criterion can be fulfilled with a wide variety of different features.

A **requirement** is a description what must be present in a product. Typically, requirements are specified in terms of the product, not in terms of the user. Requirements are Boolean: either the requirement is met or the requirement is not met. For example, a requirement may read "the mobile application must be able to sync with the client computer." It is common for a requirement to be able to be fulfilled with more than one feature. In the above example, this can be fulfilled with Bluetooth, WiFi, the cellular networks, or even a USB cable. Each of these options represents a different feature meeting the needs of the requirement. Requirements are of little use in the design process because they don't help us understand which of a set of designs better meets the user's needs.

Criteria:

Constraints in the design process through which various design alternatives are measured

A **criterion** is a constraint in the design process through which various design alternatives are measured. One criterion can typically be fulfilled by a large number of requirements, each of which can be fulfilled by a large number of features. A well-specified criteria list gives the designer latitude to pursue innovative solutions to solve the problem. For example, a criterion may be: "The user will love the way the product responds to input." From here, the designer may pursue two possible requirements: "The system must respond to user input in 0.5 seconds or faster" or "The system shall provide audible feedback with every user input." Each of these very different approaches has strengths and weaknesses. Which is better? The criterion is the metric through which the alternatives are compared.

Criterion, requirements, and features are each useful tools in various stages of the design process. Features are components of a design to help meet the needs of the stake-holders. These needs are enumerated with requirements. We evaluate the quality of a feature through criteria. Since the design process is mostly concerned about improving the quality of the product, the evaluation criteria is normally the tool of choice.

Properties of Evaluation Criteria

The criteria are designed to represent the constraints governing the design process. Therefore, the value of the criteria is a function of how useful they are in the design process. If the criteria help the designer choose between design alternatives or to refine a given design, then they are fulfilling the measure of their creation. If, on the other hand, they do not differentiate between a clearly good and clearly awful design, then they are to be hewn down and cast into the fire. Evaluation criteria have the following properties:

User centric Each criterion is described in terms of the user's needs and the user's important scenarios. When a criterion is product-, market-, or technology- centric, it is probably not a criterion at all.

Imply the variables Criteria are typically expressed directly or indirectly in terms of the variables of usability. If there is no variable implied, it is probably not a true criterion.

Measurable While not all criteria need to be measured, the better ones typically are. It should be possible to tell, intuitively and experimentally, which of a set of designs better meets a given criterion.

Identifying Criteria

The process of identifying evaluation criteria for a given persona and scenario is basically the process of converting the variables of usability into an effective design tool. There are several steps in this process: 1) articulating the variables in the context of the problem, 2) prioritizing the criteria, and 3) developing a criteria rubric. Note that each of these processes is designed to make it easier for the designer to make an accurate assessment of a design with as little work as possible.

Each of these steps will be explained in the context of the following persona and scenario for a simple web-site authentication use-case:

- Persona: Lila is a middle-age housewife with above average computer and internet skills. Lila knows how to use technology to make her life better, but is neither an early-adopter nor a fan of technology for its own sake.
- Scenario: Lila decides to do online banking. She navigates to her bank's web site for the first time and logs in.

Step 1: The Variables

The first step of the criteria-identification process is to articulate the variables of usability in the context of the persona and scenario. This is accomplished by listing each variable and expressing what each variable means in the context of the problem we are trying to solve. Note that we are not evaluating a design at this process. Back to our web-site authentication example, some criteria might be:

Variable	Criteria
Efficiency	Authentication is performed quickly
Learnability	Novice user is successful
Familiarity	Like other login screens (compatibility). Like other forms on the site (consistency).
Simplicity	N/A
Mapping	Each control should be obvious and intuitive
Motivation	Not be annoying. Be visibly pleasing
Trust	Appear reputable and authentic
Visibility	N/A

From this outline, we can derive several criteria:

- The authentication process must not take too long for any user [Efficiency].
- A novice user must be able to successfully login the first time [Learnability].
- The user should recognize all the controls and be familiar with their use [Familiarity].
- Each control's use should be obvious and intuitive to the user [Mapping].

- The authentication process must not be annoying [Motivation].
- The authentication screen should be visually pleasing to the user [Motivation].
- The appearance of the authentication controls must appear authentic and reputable so the user trusts the application more due to its presence. It must not look spoofed [Trust].

Step 2: Prioritization

Note that not all criteria are equally important for a given problem. Typically criteria are divided into three categories:

Priority 1 The priority 1 criteria need to completely capture every aspect of the design that absolutely must occur for the design to be considered a success. In other words, a good design must exemplify all the priority 1 criteria. If a design is lacking even a single priority 1 criteria, then the design must be lacking (or the criteria is misidentified?)

Priority 2 The priority 2 criteria represent all the aspects of the design that represent worthwhile endeavors. While none of the priority 2 criteria are required for the design to be a success (that is a priority 1 criteria), each represents opportunities to add value to the design. Typically we are willing to pay the price for priority 2 criteria if resources are available

Priority 3 The priority 3 criteria are good things to have but we should not go out of our way for them. Each priority 3 criteria adds value, but the gains are small and the cost may be high

Never sacrifice a priority 1 criteria for a priority 2 criteria. Our focus must always be to completely satisfy all the priority 1 criteria in every design, only looking to priority 2 criteria as time and other constraints permit.

Back to our web site authentication example, we might prioritize the criteria as:

The authentication process must not take too long for any user — Priority 2. Speed is an important consideration for any operation that must be performed many times. However, if it takes twice as long as a comparable design, that does not in itself constitute a failure.

A novice user must be able to successfully use the password field the first time — Priority 1. If the novice user is unable to authenticate, he will never come back. This constitutes a failure of the system which is the very definition of a priority 1 criteria.

The user should recognize all the controls and be familiar with their use — Priority 2. It would certainly be a benefit to the users if the controls were familiar, but that is not a prerequisite for success. It is possible an inexperienced user can figure out a novel authentication control.

Each control's use should be obvious and intuitive to the user — Priority 2. If a user had to deduce how the controls work but was able to figure it out in the end, it would be deemed a successful design. Only if the user was unable to figure out the controls would it be a failure.

The authentication process must not be annoying — Priority 3. If the user was mildly annoyed by the authentication process but was still able to complete the process quickly, it is not a failure or even too undesirable. The CAPTCHA used to identify humans when setting up a new e-mail account is a good example. However, if the user was extremely annoyed to the point of being mad, this would be raised to a priority 2 criterion.

The authentication screen should be visually pleasing to the user	Priority 3. We are always trying to make the user like our designs, but in this scenario it is not a high priority. We will be happy to achieve this criteria, but are not willing to go out of our way to achieve it.
The appearance of the authentication controls must appear authentic and reputable.	Priority 1. If the user does not trust any aspect of an online banking experience, we failed. Thus anything that appears amateurish, non-standard, or "questionable" to any capacity could severely decrease the overall usability of the entire site.

Step 3: Rubric

Now that the evaluation criteria are identified for a given problem, one final step needs to be taken to make them truly useful for design activities. This final step arises from the fact that the evaluation criteria tell us what the design needs to be, but not how well it meets those criteria. To fill this gap, we often need a rubric.

A rubric is a quantification tool designed to facilitate identification of a score from qualitative data according to pre-established criteria. Rubrics are commonly used in academic settings as a tool to communicate to students the quality of work that is asked of them and as an instrument by teachers to grade student work. Similarly, rubrics can be used to help designers evaluate different iterations of their designs by providing pre-established benchmarks for design quality.

There are three components to a rubric. The first is the rows, representing individual criteria by which something is to be evaluated. The second is the columns, representing degrees of quality or benchmarks for each row. The final component is the weighing scale, indicating the relative degree of importance for each row. Note that we only use priority 1 and priority 2 criteria for our rubric. This is because "we should not go out of our way for [Priority 3 Criteria]".

Each row maps directly to an evaluation criteria. Typically as a row header we identify a one or two word title to represent the criterion.

Next, each cell in the row represents a degree of quality for each criterion. It is common to label each column as Exceptional, Good, Acceptable, Flawed, and Horrible. Recall the scale provided with each variable of usability. Five benchmarks were provided: the neutral benchmark (score 0), the failure benchmark (score -2), the run-away success benchmark (score 2), and a couple scores in between. Since criteria are derived from the variables of usability, it follows that an evaluation criteria rubric should be an adaption of the scales for each variable. For example, consider the criterion for the authentication example: "The authentication process must not take too long for any user."

Consider also the efficiency scale from the variables of usability:

Score	Description
2	Users of the interface are more productive than with any comparable design.
1	Overall user performance is better than most similar designs.
0	Productivity of members of the target user group is typical when compared to similar designs.
-1	Overall user performance with the interface is below average or some task yields the worst performance possible.
-2	Users of the interface are less productive than with any comparable system.

An example of adaptation of the above variable scale to our criterion is:

Score	Description
Exceptional: 2	Users can authenticate faster than with any comparable design.
Good: 1	Users can authenticate faster than with most similar designs.
Acceptable: 0	Authenticate speed is average.
Flawed: -1	Authentication speed is below average or occasionally is the worst.
Horrible: -2	Users authenticate slower than with any comparable system.

The final part of the rubric is the weighing scale. This is to represent the criteria prioritization we established earlier. In order to represent the prioritization completely, it is necessary to weigh the important (Priority 1 items) more heavily than the rest (Priority 2), and to represent that all the priority 1 criteria must be met. The first consideration can be readily achieved by giving the priority 1 criteria two or three times the weight of the rest. The second is rather tricky. How do you indicate that none of the priority 1 criteria have a final score less than zero?

The following is a rubric representing the evaluation criteria identified for our online banking authentication example:

	Exceptional 2	Good 1	Acceptable 0	Flawed -1	Horrible -2
Novice 40%	Virtually all novice users will become experts	Most novice users will become proficient	All novice users will eventually be able to authenticate	Many novice users will be able to authenticate	No novice user will ever figure this out
Authentic 30%	Users are convinced it is authentic	Inspires confidence	Nothing encourages or discourages feelings of authenticity	Some will suspect that this is fake	Everyone will think this is fake
Time 10%	Fastest design known	Faster than most similar designs	Authentication time is average	Slower than most	Takes longer than with any comparable system
Familiar 10%	The interface is indistinguishable from others	Elements of the design are shared with other designs	The design is novel	Previous experience will be a disservice	Previous experience will be a severe handicap
Obvious 10%	Most users will figure out the design	Design cues will lead the user to understanding	No particular mental model is enforced	Misconceptions will not be discouraged	Intuition will mislead the user

There are a few things to observe about the rubric. First, consideration must be made that each cell in the rubric has a textual description that accurately describes the quality of the corresponding criterion (mentioned at the row header) at the corresponding quality level (mentioned at the column header). It is useful to look at each cell and ask yourself "is that textual description what I meant?"

The second thing is that, for a given row or criterion, the textual description needs to be unambiguous. If another evaluator were to judge a design based on this rubric, would he make the same conclusions and compute the same score?

The final thing to take into account when making a rubric is that the creation process is hard! Expect to make mistakes the first few times. It is worthwhile to make a draft and evaluate a few existing products before using it with the design process. As an experienced rubric author in the academic setting, I can testify that there are always problems with the first draft!

Example

Consider the mountain bike team web site used for the previous two chapters. Our personas are Rachel, Morgan, and Amber

Figure 06.1 - Rachel

Rachel is a 42 year-old mother of 4. Her second oldest is Morgan, the only member of the team in her family. While she is not "tech savvy," she is not ignorant of technology either. Rachel texts with her older children, has a Facebook account, and regularly shops online. In fact, her husband just bought her an iPad and she has reluctantly started bringing it with her on her daily routine.

Rachel is constantly trying to coordinate the activities of her various family members. This is particularly challenging with her daughter Morgan because her races are out of town every weekend. To lighten her load, Rachel works with two other parents to carpool and share making meals.

Figure 06.2 - Morgan

Morgan is a 17 year-old student a Madison High School. She is not a big fan of organized sports; why watch other people have fun when you could be doing it yourself? Many of Morgan's friends are also on her team but quite a few are not. They don't really understand mountain biking and why she is gone so many weekends in the fall. Morgan wants her friend to know about the latest race, but does not want to re-post all the pictures on her social media. After all, not everyone cares about her race! Morgan is a typical millennial with a high degree of web and computer literacy.

Figure 06.3 - Amber

Amber is a friend of Morgan and has even considered joining the team. It looks like they have so much fun, but she is just not sure that the team is for her. She likes riding her bike and even mountain biking, but racing is a totally different level of commitment. When Morgan posts a picture on social media of her latest race, Amber goes straight to the team web site to see how it went. It looks like so much fun!

The first scenario is Rachel sitting in her kitchen talking with her family:

Rachel is preparing dinner when her children start coming home from school. While eating a muffin, Morgan asks her about the upcoming weekend. She wants to know who is driving and what they will be eating. Rachel walks over to her laptop on the dining room table and clicks on the Mountain Bike Team web site bookmark. Seeing the top-level navigation control, she clicks on the "Schedule" tab and quickly finds the next race. As she mentions a few details to Morgan, she looks for any changes or notes from the coach. There are none!

As she talks with Morgan about the carpool plan (Rachel is driving this weekend) and the food (pasta again!), Rachel idly navigates to the "Picture Exchange" tab. She notices that some new pictures were added to the top of the list. As they firm up their meal plan, Rachel periodically calls Morgan over to look through some of the pictures. A few other curious family members peer over their shoulders as they laugh. Perhaps Morgan's brother will ride next year after all!

The scenario for Morgan and Amber is combined:

Morgan is over at Amber's house to work on homework together. As they get stuck on a tough problem, the conversation immediately goes to more interesting things. Amber asks Morgan about how her race went over the weekend. Rather than just tell her, Morgan jumps to the team web site and clicks on the pictures for the latest race. As they run through the pictures, Morgan tells story after story about their misadventures. One particular picture has them both in stiches. Several stories later, no one is thinking about math anymore.

To identify the criteria, we need to identify the relevant variables, prioritize the criteria, and then create a rubric.

Step 1: The Variables

Variable	Criteria
Efficiency	Rachel should be able to quickly find details about the upcoming race. Rachel should be able to find pictures of her daughter without much difficulty. Discovering notifications should be very easy for Rachel.
Learnability	Rachel will be able to figure out all aspects of the site in just a minute or two.
Familiarity	The site layout and controls should be similar to other sites that Rachel, Morgan, and Amber uses.
Simplicity	Mental model not larger or more complex than a simple newsletter.
Mapping	Rachel, Morgan, and Amber should never get lost. The site should accurately represent what the team is all about.
Motivation	The design and the layout of the site should draw Rachel, Morgan, and Nicole in.
Trust	Rachel never feels like she needs to hunt around to find what she is looking for.
Visibility	Rachel, Morgan, and Amber should never get lost. Important information is very prominent.

Step 2: Prioritization

Based on the scenario and the various use-cases that were developed, the following prioritized evaluation criteria might be used:

Priority 1

- Important information is very prominent, easy to find, and does not require hunting [Trust, Visibility, Efficiency]
- The design and the layout of the site must draw Rachel, Morgan, and Amber in [Motivation]

Any design that fails to meet these two criteria is a failure, regardless of any other qualities it may have. To verify this, imagine two situations. First, Rachel thinks she has the race information figured out and then, when she arrives, realizes that it was moved to next week! Clearly this is unacceptable, so the criteria is a priority 1.

Second, imagine a site that is boring and bland. Why not just use e-mail to get the same results? If the site fails to draw in new recruits and if people use other means to communicate, then the site will fail to reach the measure of its creation. Since this is also unacceptable, the second criteria is a priority 1.

Priority 2

- Rachel and Morgan will be able to figure out all aspects of the site in just a minute or two [Learnability]
- The site should accurately represent what the team is all about [Mapping]
- The site layout and controls should be similar to other sites that Rachel, Morgan, and Amber uses [Familiarity]
- Rachel, Morgan, and Amber should never get lost [Mapping, Visibility]

Each of these criteria adds value to the product. So much value, in fact, that we are willing to make significant sacrifices to achieve them. Note that we do not have to achieve any of these criteria to have a successful product; if that were the case these would be classified as Priority 1.

Priority 3
- Mental model not larger or more complex than a simple newsletter [Simplicity]

The final set of criteria is called "icing on the cake." In other words, each is nice to have but does not directly correspond to value to the user. We will be happy to achieve each of these, but they are not worth going out of the way for.

Step 3: Rubric

The final step is to convert these criteria to a single rubric to facilitate rapid and accurate evaluation of a given design or of several design alternatives. Because we have so many compelling Priority 2 criteria, we can safely discard all the Priority 3 for this rubric.

	Exceptional 2	Good 1	Acceptable 0	Flawed -1	Horrible -2
Important Info 40%	Rachel is nearly certain to not miss important information	The important info is prominently displayed	It is clear where the important information is posted	One can find the important information with effort	It is very difficult to find important information
Motivating 30%	The site draws people in and encourages sharing	The site is pleasing to the eye and interesting	Everything is professional	The site is boring	Unprofessional, ugly, or punishing
Learnable 10%	Everyone figures it out immediately	Most users will figure out the site in just a minute or two	Familiar and learnable, but more difficult than necessary	Some users will never discover some features of the site	Many users never discover some features of the site
Representative 10%	The site captures the culture and feeling of the team	The site resonates with members of the team	The data in the site is accurate and trustworthy	Some aspects of the site are misleading or incorrect	The site misleads visitors as to what the team is all about
Understandable 10%	Extremely simple and straightforward	Most users would call the site "simple"	As one would expect	More complex than necessary	Requires much more work

Comprehension check

Quiz 1 - 3 **Priority 1,2,3**

In a sentence or two, define a priority 1, 2, and 3 criteria.

Quiz 4-8 **Categorization**

Categorize the following examples as a feature, requirement, or criterion:

- The user must be able to type the filename in the edit control
- The file can be printed to a PostScript printer
- The user must notice the [File] button in 2-3 seconds
- The project must be completed in two weeks
- The user should like the way the phone feels in her hand

Quiz 9 **Priority 1**

When is it acceptable to sacrifice a Priority 1 criterion for a Priority 2 criterion?

Quiz 10 **Personal goals**

Please list your personal goals for reading this book. Be reflective, this should take a bit of time. Also, be honest. "Because my teacher made me" could be a valid, honest goal.

Problems

Problem 1

Figure 06.4 - CB Radio

CB radio

Identify evaluation criteria for the CB radio, using the persona and scenario developed earlier. Please follow the following steps:

1. List the eight variables of usability
2. For each variable, list a couple criteria

It is not necessary to prioritize this criteria list.

Problem 2

Figure 06.5 - Cordless drill

Cordless drill

Identify the evaluation criteria for a cordless drill. Again, use the persona and the scenario developed from a previous lesson. Please...

1. Articulate each of the variables in the context of the cordless drill, the persona, and the scenario
2. Prioritize each criteria (1, 2, or 3)
3. Create between five and ten criteria that capture the essential characteristics of the cordless drill user interface design
4. Make a rubric from the criteria

Problem 3

deep dive 96
dropdown listbox 174
edit control 175
efficiency 22, 114, 159, 165
elevator pitch 85
enclosing 129
ergonomics 5
evaluation 13, 14
evaluation criteria 71, 82, 93, 97, 115, 116, 118, 119, 124, 126, 189
execute sequence 16
execution 13
executive summary 85, 97, 118, 126

Figure 06.6 - Index feature

Index feature

As with problem 2, identify the evaluation criteria for an index feature for an electronic textbook. When finished, please list five to ten criteria and provide a rubric

Problem 4

Figure 06.7 - Stopwatch

Stopwatch app

As with problem 3, identify the evaluation criteria for a stopwatch mobile application. When finished, please list five to ten criteria and provide a rubric

07 Heuristic Analysis

Analysis is the process of learning about a given problem in a systematic way. This can include learning about a problem domain before beginning the design process or evaluating the quality of an existing design.

Three Analysis Techniques

Expert	An expert gives it a once-over
Experimental	See how it works with real users
Heuristic	Component analysis

There are three basic types of analysis techniques: expert analysis, experimental analysis, and heuristic analysis. Most studies consist of a combination of the three, but it is instructive to look at them individually.

Expert analysis consists of a known expert giving the design an once-over. Consider, for example, a team manager analyzing a new race. An expert analysis would consist of the driver (the expert) giving the car a few turns around the circuit. The kind of feedback the driver might give include: "The engine is hesitant at the low RPMs," "The steering feels sloppy in the chicane," and "High-end acceleration is excellent." Expert analysis can often be difficult to obtain because an expert must be on hand. Similarly, expert analysis can be difficult to interpret. Two different experts are likely to give you two different answers!

Experimental analysis consists of observing how the system performs as a unit. Back to our race car example, an experimental analysis would consist of comparing lap times of the new car against last year's model. From this, we can see which completes a faster overall lap time, which has a higher top-end speed, and which can take a given corner faster. In many ways, this is the most authentic evaluation technique. After all, we are directly measuring what matters most: speed. However, it is often very difficult to answer "why" questions.

Heuristic analysis consists of studying the pieces of the design. In the race car example, this would consist of studying the drag coefficient of the car, the horsepower and torque of the engine, the center of gravity, and the braking capacity. Since these variables are known to influence the performance of the car, they are key indicators as to how fast the car will be. Heuristic evaluation data tends to be easy to interpret: the variables tell you exactly what is good or bad about a given design. Unfortunately, they tend to be difficult to extrapolate. If the variables do not represent all of the key performance indicators, they may not paint a complete picture of the overall design.

Figure 07.1 - Race car

Steps to Heuristic Analysis

While all three analysis methods are commonly used in HCI problems, heuristic analysis is perhaps used most often in the early phases of a project. There are six steps to the heuristic analysis process:

1. Identify the target audience The first and most important step in any HCI activity is to understand the target audience. Without knowing the users for whom the system is intended, it is impossible to perform a meaningful analysis.

2. Identify the user's tasks Once the target audience is properly understood, one must then understand the task the users are attempting to accomplish with the system.

3. Identify the criteria How can one distinguish success from failure in a given design?

4. Individual analysis For each criteria, conduct a detailed analysis. Try to discover all that is right or wrong about the design from that criteria's perspective.

5. Criteria hierarchy Map the relationship between the criteria. Determine if a problem stemming from one criteria can impact others.

6. Draw conclusions From your individual analysis results taking into account the ranking or weighting performed previously, generate a holistic overview of the usability of the system.

1. Identify the Target Audience

Recall that HCI is essentially a communication problem. One half of this communication is the system and the other is the user. Without a detailed understanding of the target audience, it is impossible to correctly characterize this communication.

Chapter 04 describes the process of identifying and characterizing the target audience in more detail. For now, it is important to remember that data collection is an important and ongoing process. Also, the best way to characterize and communicate complex data about a diverse target audience is through one or more personas.

2. Identify the User's Task

Usually performed in tandem with the identification of the target audience, user task identification is the process of describing all aspects of the interaction framework. This can be accomplished with a set of use cases, user stories, or scenarios. Regardless of the representation chosen, it is essential that the usability engineer(s) know what the users are going to do when they interact with the system.

3. Identify the Criteria

The next step in any heuristic analysis is to determine the key performance indicators. For many situations, this can be the most difficult part. What exactly makes for a great employee, cell-phone, or English essay? Fortunately, in the context of Human-Computer Interaction, we have two useful tools: the variables of usability and the evaluation criteria.

The variables of usability is an excellent choice for heuristic analysis if only a small number of designs are to be evaluated. While the variables are designed to be useful in all usability contexts, they are generic and therefore do not take the specifics of the problem into account. In other words, not all the variables are relevant to a given usability problem. One must therefore determine which variables influence the overall usability picture, and which do not. In the typical usability situation, one or two variables can be excluded. Be cautious if you find yourself removing most of them!

In most scenarios, each variable will have a different degree of influence or weight on the overall usability picture of the problem. The usability engineer will need to make an informed decision as to which variables are most relevant in the context of the problem. There are three ways this is commonly done: a simple ordering, a grouping, or a weighing. The simple ordering involves ranking the variables from most important to least. One advantage is how easily and quickly that can be done. However, if two variables are very important and the rest are not, then a simple ranking will not capture this relationship. Grouping, on the other hand, is designed to capture just such associations. A typical example might consist of three groups: very important, somewhat relevant, and little influence. The final method is to create a weighted average where the percentage of the overall impact is enumerated for each variable. For example, efficiency and motivation may constitute 30% apiece, learnability, familiarity, and mapping may constitute 10% apiece, and the balance may fill in the rest. While this is certainly the most accurate and powerful model of the three, it forces the usability engineer to quantify relationships using insufficient data. As a result, the numbers often feel "made up."

A second method is to use the evaluation criteria. Here (please see Chapter 06 for more detail), the usability engineer identifies evaluation criteria based on the variables of usability and the specific needs of the problem being analyzed. The end result may be prioritized criteria or a fully flushed out rubric. These criteria take time to develop, of course, but the resulting criteria facilitate fast and accurate analysis

Which method should you use? As a rule of thumb, one-off or ad-hoc analysis is better served using the variables of usability. The cost of identifying the evaluation criteria or rubric is simply too high. On the other hand, if multiple designs are to be analyzed as is the case in the design phase of a project, then the up-front cost of defining evaluation criteria is worth it.

4. Individual Analysis

Individual analysis is accomplished by conducting an analysis, one criteria at a time. The usability engineer attempts to make an exhaustive list of all the interface design flaws for a given criterion and then repeat the process for all the criteria. By focusing analysis efforts on a single criterion at the exclusion of all others, it is possible to find errors that may be masked by one glaring error represented by another criterion. Of course, more effort should be spent on higher priority criteria as identified in the third step.

There are two main results from this process: a list of defects and a score. The list of defects is useful for making improvement suggestions. The score is useful for comparing the relative merits for competing designs.

5. Criteria Hierarchy

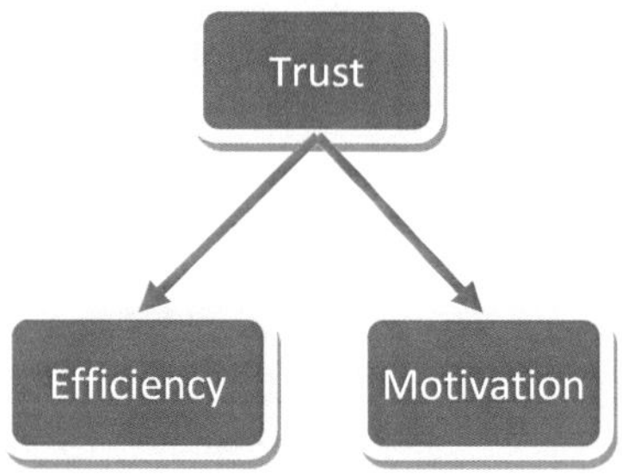

Figure 07.2
Usability variables are often related hierarchically. These relationships can be represeted with a simple tree graph.

The next step is to collect all the individual findings uncovered in the preceding steps into a single coherent report. As with the individual analysis, there are two main parts of a usability report: a list of defects and an overall score.

While the defect list is compiled almost directly from the individual analysis, additional thought may be required. The first item to look for is defects spanning multiple variables. In these cases it is usually best to report them as a single defect. Another item to look for is when there is a hierarchical relationship between defects. For example, breakdowns in trust commonly yield efficiency and motivation problems. Here it would be appropriate to report the root cause as well as all the symptoms and side effects as a single defect. Illustrating the relationship with a simple graph is often instructive as well. Usability variables are often related hierarchically.

6. Draw Conclusions

The final component of a usability report is to draw conclusions. This can be qualitative or quantitate, depending on the needs of the analysis.

Qualitative, as you may recall, involves using words and descriptors. An example of a qualitative report may be: "The preponderance of extra clicks results in a not only a severe problem with efficiency, but also motivation. The user is highly agitated every time an unnecessary prompt or screen is presented." Another example might be: "The users are delighted by the entire user experience. This product is likely to be a resounding success when it is released." Qualitative reports are most applicable in the preliminary analysis phase of product development (before the first design is drawn) and in the verification phase (when we are trying to determine the list of defects or ascertain the overall quality level).

Quantitative, on the other hand, involves using numbers and statistics. Here we try to use a single number or collection of numbers to represent the quality of a given design. Typically the number itself holds little meaning. It is far more illustrative to discover which component of the design is dragging down (or up) the score. One counter-example is during the design process. As a designer is iterating through multiple design alternatives or is seeking for ways to improve the design, then the heuristic score is commonly used as a benchmark to measure progress or to find direction. The simplest way to compute the score is by summing the product of the individual scores (by variable: $S_{variable}$ for the score of the variable) with their weight ($W_{variable}$ for the weight for the variable):

$$\text{Score} = \text{S}_{\text{efficiency}} \times \text{W}_{\text{efficiency}} + \text{S}_{\text{learnability}} \times \text{W}_{\text{learnability}} + \ \dots$$

Another common way to compute the score is by using a rubric. A rubric is a tool used to quantify (reduce to numbers) inherently qualitative data. In other words, it helps us to compute scores based on a collection of criteria. You may notice that rubrics are commonly used by teachers to score complex assignments: each row corresponds to an objective or criteria by which the student work is evaluated and each column corresponds to a quality level of the work. Similarly, rubrics can be used to score designs. Consider the following rubric based on the variables of usability:

	2	1	0	-1	-2	Weight	Score
E	Most productive known	Better than most	Comparable	Below average	Less productive	.1	-0.2
L	Almost certain to be an expert	Some will be expert	Most will achieve intermediate	Few will do better than intermediate	User will remain a novice	.1	0.0
F	Indistinguishable	Works like other designs	Novel	Previous experience will mislead	Previous experience will handicap		0.0
S	Smallest and simplest known M.M.	M.M. smaller than most	Most will be able to internalize	Understandable	Exceeds capacity to understand	.2	-0.2
Ma	Highly likely will form a valid M.M.	Encourages a valid M.M.	Does not encourage any specific M.M.	Suggests invalid M.M.	Discourages from forming valid M.M.	.3	0.3
Mo	Satisfaction, pleasure, etc.	Likely to appeal	No emotional response	Prefer to not use the interface	Distaste, irritation, annoyance		0.0
T	Convinced they have complete control	Relaxed and stress-free	System is predictable	Hesitant	Nervous, distrustful	.3	-0.3
V	Most convenient known format	Convenient	Can be found	Significant searching	Cannot be found		0.0
							-0.4

Thus, according to the weighting and individual scores of this example, the overall score is -0.4 or "somewhat below average."

Example

To illustrate this process end-to-end, consider a coat closet door handle situated near the front of a house. The user of focus will be a guest to the house wishing to put his coat in the closet during a visit. One interesting thing to note about this example is that the handles do not turn; you open the door by pulling to overcome a friction lock.

Figure 07.3 - Closeup of door handles of the non-twisting variety

Step 1: Identify the Target Audience

There are two classes of users here: the inhabitance of the house and the visitors. Since this door is for the coat closet, the visitor is the most important segment of the target audience.

Step 2: Identify the User's Task

The task is to open the door. The entire scenario involves a few important details. First, notice that the user will use only one had because the other is holding his or her coat. The second is that the user is distracted, probably talking with the host as the coat is being put away. Finally, there is a bit of urgency: the user does not want to be embarrassed about having difficulty opening the door or hanging the coat.

Step 3: Identify the Criteria

In this example, all the variables are relevant except simplicity. The system is so straightforward that there is little the interface can do to complicate or simplify the user's mental model.

The two most important variables in this scenario are learnability and motivation. Learnability is important because the target user is a novice and he will need to figure out the interface with just about every use. Motivation is also important because the interface occupies a prominent location in the house and needs to be visually pleasing. The weight for this group of variables is 30%.

The next most important variable is familiarity. There is a strong precedent for door handle designs, a precedent that could work either to the designer's benefit or to his disadvantage. The weight for this variable is 20% because it is nearly as important as the first group.

The final tier of variables includes efficiency, mapping, trust, and visibility. Each plays a role in the overall interface, but the role is minor. The final 20% of the score will be divided evenly among them.

Step 4: Individual Analysis

The results of the individual analysis are the following:

- **Efficiency** -0.5 The door sticks and requires more effort than one would expect.
- **Learnability** 2.0 Everyone masters the door with effort.
- **Familiarity** -1.0 This door is inconsistent with the other doors in the house: The closet handle does not turn whereas the bathroom door handle does. They both look the same.
- **Simplicity** 2.0 Exceedingly small mental model.
- **Mapping** -1.5 Handles are meant for turning. The affordance is misleading.
- **Motivation** 2.0 Most guests agree the door is aesthetically pleasing and comfortable to the touch.
- **Trust** -1.0 After one attempt, most guests hesitate before using other doors in the house.
- **Visibility** -1.0 No visual clues distinguish a turning handle from a non-turning handle. A critical piece of information is missing.

Step 5: Criteria Hierarchy

The largest problem with the door handle is that it does not turn as the user expects. This expectation is the result of is inconsistency with other door handles (familiarity) as well as a violation of the handle affordance (mapping). Note that if the lack of a twisting operation was more obvious (visibility), the mapping issue might be mitigated. The end result is a decrease in trust.

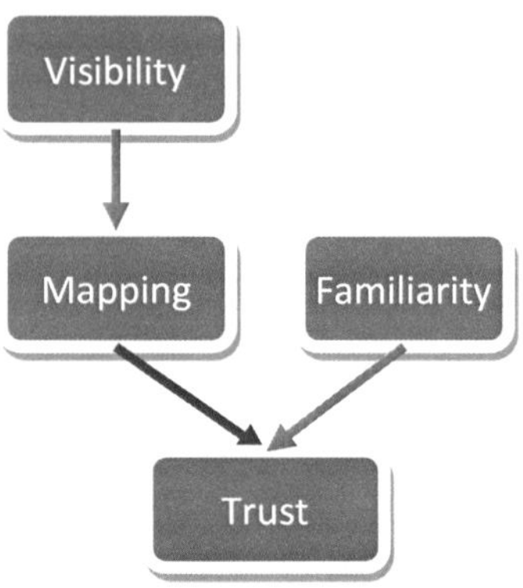

Figure 07.4 - The relationship between visibility, mapping, familiarity, and trust in the door handle example

Step 6: Draw Conclusions

The second problem is unrelated to the first. The friction locking mechanism is too tight, resulting in unnecessary difficulty in opening the door. This results in a breakdown in efficiency. The overall score for the door handle is represented below:

	2	1	0	-1	-2	Weight	Score
E	Most productive known	Better than most	Comparable	Below average	Less productive	.05	-0.025
L	Almost certain to be an expert	Some will be expert	Most will achieve intermediate	Few will do better than intermediate	User will remain a novice	.30	0.60
F	Indistinguishable	Works like other designs	Novel	Previous experience will mislead	Previous experience will handicap	.20	-0.20
S	Smallest and simplest known M.M.	M.M. smaller than most	Most will be able to internalize	Understandable	Exceeds capacity to understand		0.00
Ma	Highly likely will form a valid M.M.	Encourages a valid M.M.	Does not encourage any specific M.M.	Suggests invalid M.M.	Discourages from forming valid M.M.	.05	-0.075
Mo	Satisfaction, pleasure, etc.	Likely to appeal	No emotional response	Prefer to not use the interface	Distaste, irritation, annoyance	.30	0.60
T	Convinced they have complete control	Relaxed and stress-free	System is predictable	Hesitant	Nervous, distrustful	.05	-0.05
V	Most convenient known format	Convenient	Can be found	Significant searching	Cannot be found	.05	-0.05
							0.8

Observe that the overall score of the door handle is actually quite high despite the obvious problems. This is because the design scored well on the areas that matter most (Learnability & Motivation), more than offsetting the areas that didn't (Familiarity, Mapping, etc.).

Comprehension check

Quiz 1 **Terms & Definitions**

For each of the following description, name the corresponding evaluation type

- The evaluation technique that does not answer "why" questions
- A study of the pieces of the design
- The evaluation technique that is easiest to interpret
- A direct measurement of the design
- The most authentic evaluation technique
- The most difficult form of evaluation to interpret
- An expert opinion

Quiz 2 **Steps in a Heuristic Evaluation**

Order the following steps to a heuristic evaluation:

- What are the important metrics?
- How do the criteria relate to each other?
- Who is this interface meant for?
- What is good and bad about the design?
- What are the users trying to accomplish?
- Look at each criteria in isolation of the rest

Quiz 3 **Evaluation Techniques**

For each of the following scenarios, choose the best evaluation technique:

- At the end of the product cycle, we would like to find out what percentage of the users will actually use a few feature that was added to the product
- While working in the design process, we wish to identify the strengths of and weaknesses of the design

Quiz 4 **Compute Score**

Consider the following weighting values (E:20% L:30% F:10% S:10% Ma:20% Mo:0% T:0% V:10%) and scores (E:0, L:2, F:-1 S:-1.5, M:-1.5 M:2, T:-2, V:1.5), what is the overall usability score computed to 1 decimal of accuracy?:

Problems

Problem 1

Figure 07.5 – Infotainment system

Infotainment system

Perform a full heuristic analysis for an "all glass" instrument cluster for a modern car. All the controls are reachable through a single touch interface. This includes identification of the target audience, a task analysis, identification of the evaluation criteria, individual analysis, criteria hierarchy, and finally draw conclusions.

UNIT THREE 3

Design

Design is the process of systematically searching for solutions to interface challenges so the needs of the target audience can be met

08 Executive Summary

Many believe the source of uninspired or off-base designs result from lack of talent, imagination, or experience. It turns out, however that the greatest source of error in the design process is a lack of focus. It is so easy to get caught up in the technology that we forget the purpose of the feature.

To illustrate this point, consider a small development team consisting of a developer, a designer, a marketer, and a manager. The manager charges this team to develop a new web site for the local high school mountain bike team. As he gives this charge, the manager envisions an incremental improvement over the current design that fixes several of the more obvious flaws. The marketer who wants the team to grow, is visualizing a snappy new feature that demos well and makes for a great screen shot. The developer thinks database technology is the answer and wants to expose a query-type interface to the users. Finally the designer is trying to reconcile many different approaches into a single design and is having trouble coming up with something workable. This problem is exasperated every time the small team meets. As the feature is discussed in general terms, each member visualizes the discussion in the context of his idea of how the design should look. These different visions get solidified with time and get coded into various aspects of the product. The product begins to feel disjoint and the team meetings become contentious; it is readily apparent that the project is in serious difficulty.

On the surface, it may seem impossible for such a small team to suffer from such a communication problem. However, this problem is actually quite common. It is very difficult to keep all members of a team on the same page, regardless of the team size. Clearly, what is needed here is some tool to keep everyone on the same page. The most effective tool at our disposal to do this is the executive summary.

Definition

Every successful organization you can think of, be that commercial or eternal, has a mission statement. This statement encapsulates the main purpose or philosophy of the organization. While many believe the mission statement to be hollow words having no purpose outside the occasional pep rally, the true value is much more subtle. If internalized by the member of the organization and frequently referenced, the mission statement can be a powerful tool to help everyone focus on what is important.

Executive Summary:

A short description of what the feature or product is all about

The executive summary for a feature and the mission statement for an organization are exactly the same thing. For example, the mission Disney is "to make people happy." Therefore, every single activity performed by Disney should be directed to this purpose. This should direct the management of theme parks, the merchandise, and media production. Whenever a designer, worker, or manager needs to make a decision anywhere in the organization, they simply need to ask themselves "what will make people happier?"

The executive summary should completely encapsulate the essence of the feature. When you describe the feature to someone, they need to walk away knowing exactly what the feature is about. This, of course, is extremely difficult to do: it requires you the designer to have clarity and focus. It is difficult to fake an executive summary.

Characteristics

The characteristics of an executive summary include:

Completeness The executive summary must completely describe the feature. All the big decisions, directions, and components need to be included

Conciseness The executive summary must be very short: on the order of a couple sentences. There is no room for fluff or buzzwords in an executive summary

Convincing The reader needs to see the value of both the project and the approach

Repeatable The executive summary needs to be memorable and repeatable. If it cannot be easily internalized, its effectiveness as a communication tool is diminished

The convincing characteristic of the executive summary is perhaps the most difficult. How can you convince someone that the product or feature is worthwhile? The answer, of course, is that the product or feature must be worthwhile. It must be compelling. It must have a clear value proposition.

Value Proposition:

The rational why a customer or user will be compelled to use a product

A value proposition is the rationale why a customer or user will be compelled to use a product. If the value proposition is clear, the convincing characteristic of the executive summary will be easy to achieve. If it is not, then the members of the design team (and the customer and user for that matter) will be left wondering if the feature is worth the effort. Discovering, refining, and boosting the value proposition is a key process in any design activity

Examples

Examples of highly successful mission statements:

> *At Microsoft, our mission and values are to help people and businesses throughout the world realize their full potential (Microsoft Corp.)*
>
> *To organize the world's information and make it universally accessible and useful (Google)*
>
> *We save people money so they can live better (Wal-Mart)*
>
> *To make people happy (Walt Disney)*
>
> *Enhancing the lives of those we touch by helping people reach their goals (Melaleuca)*
>
> *To enable all young people, especially those who need us most, to reach their full potential as productive, caring, and responsible citizens (Boys and Girls Clubs of America)*

> *Dedicated to creating the most epic entertainment experiences... ever! (Blizzard)*
>
> *To bring inspiration and innovation to every athlete in the world (Nike)*

The Bible even has a few examples of personal mission statements:

> *My soul doth magnify the Lord (Luke 1:46)*
>
> *Choose you this day whom ye will serve ... but as for me and my house, we will serve the Lord (Joshua 24:15)*
>
> *The Lord is my defense; and my God is the rock of my refuge (Psalms 94:22)*

Examples of an executive summary for a software product:

> *Office 2000 bridges the gap between the World Wide Web and desktop productivity software (Microsoft Corp.)*
>
> *The PDF feature enables users to take static versions of their documents anywhere without having to worry about compatibility or document reflow. (Microsoft Corp.)*

Creating an Executive summary

The executive summary is the first step in the design process. When you first begin the design process, it reflects your first impressions as to how you will go about solving the problem. It reflects not only what you think will be good about your design (the value proposition), but also what the key characteristics are likely to be.

The executive summary is the last step of the design process. After you have worked through many designs, built endless prototypes, gathered feedback from real and convenient users, and worked through all the details, it represents the summation of your efforts. In this respect, it reflects your big decisions and what makes your final design different than all your original ideas.

The executive summary should always reflect the current thinking of the designer. It is a work in progress, subject to alteration as new evidence presents itself. In other words, it is never set in stone.

There is no formula for writing an executive summary. Instead, the best way to write one is orally and iteratively. Start by explaining to a friend what the product is all about. Discuss why the product was started in the first place, who the product is for, and some of the major design decisions. Include in this discussion the value proposition. Your friend who is patiently listening to you describe this feature will probably have many questions. As thorough as you think you are being, you almost certainly will leave out some of the key items. At times he will be confused, at times you will be frustrated. This is natural; our first pass at describing something as complex as a software feature is probably very rough.

When this half-hour long process is over, find another friend and do the same thing. Learning from your mistakes from the first explanation, you will probably do a much better job; less questions are likely to result and fewer points will have to be rephrased. This is because your understanding of the feature is being clarified with each explanation. As this process continues through several friends, the time it takes to describe the feature will drop from a half-hour to just a few minutes.

This is possibly best explained by an analogy. River rocks are always smooth and round. Of course they do not start out that way: initially they were irregular and rough. However, as they spend thousands of years in stream-beds, they tumble and rub against each other. The edges are the first to be worn away because they stick out the most. After a long time, no edges remain. The same process works when we verbally explain something to another person. With practice, the rough edges get more attention and become smooth until the entire explanation is as polished as a river rock.

Elevator Pitch:

Relating the essence of a feature to someone in thirty seconds with no visual aids

When you can explain your feature to a stranger in thirty seconds without visual aids, your executive summary is ready for writing. We call this the **elevator pitch**: relating the essence of a feature to someone while riding in an elevator.

Example

Recall the scripture navigation problem we have been working on in the past few chapters. We have identified the two primary personas:

Figure 08.1 - Rachel

Rachel is a 42 year-old mother of 4. Her second oldest is Morgan, the only member of the team in her family. While she is not "tech savvy," she is not ignorant of technology either. Rachel texts with her older children, has a Facebook account, and regularly shops online. In fact, her husband just bought her an iPad and she has reluctantly started bringing it with her on her daily routine. Rachel is constantly trying to coordinate the activities of her various family members. This is particularly challenging with her daughter Morgan because her races are out of town every weekend. To lighten her load, Rachel works with two other parents to carpool and share making meals

Our first draft will be for our student athlete persona. We will call her Morgan.

Figure 08.2 - Morgan

Morgan is a 17 year-old student a Madison High School. She is not a big fan of organized sports; why watch other people have fun when you could be doing it yourself? Many of Morgan's friends are also on her team but quite a few are not. They don't really understand mountain biking and why she is gone so many weekends in the fall. Morgan wants her friend to know about the latest race, but does not want to re-post all the pictures on her social media. After all, not everyone cares about her race! Morgan is a typical millennial with a high degree of web and computer literacy.

We identified two scenarios:

Rachel wants to know some details about the upcoming race. As she mentions a few details to Morgan, she looks for any changes or notes from the coach. There are none! Rachel notices that some new pictures were added to the top of the list. As they firm up their plan, Rachel periodically calls Morgan over to look through some of the pictures. A few other curious family members peer over their shoulders as they laugh.

Morgan is over at Amber's house to work on homework together. As they get stuck on a tough problem, the conversation immediately goes to more interesting things. Amber asks Morgan about how her race went over the weekend. Rather than just tell her, Morgan jumps to the team web site and clicks on the pictures for the latest race. As they run through the pictures, Morgan tells story after story about their misadventures. One particular picture has them both in stiches. Several stories later, no one is thinking about math anymore

We have also identified the metrics by which we can evaluate various design alternatives:

Priority 1	• Important information is very prominent, easy to find, and does not require hunting [Trust, Visibility, Efficiency] • The design and the layout of the site must draw Rachel, Morgan, and Nicole in [Motivation]
Priority 2	• Rachel and Morgan will be able to figure out all aspects of the site in just a minute or two [Learnability] • The site should accurately represent what the team is all about [Mapping] • The site layout and controls should be similar to other sites that Rachel, Morgan, and Amber uses [Familiarity] • Rachel, Morgan, and Amber should never get lost [Mapping, Visibility]
Priority 3	• Mental model not larger or more complex than a simple newsletter [Simplicity]

First draft

The first attempt at the executive summary is to just try to get all the ideas down into a single, poorly organized brain-dump:

The mountain bike team web site will be one place where all members of the team can find what they are looking for. It will contain pictures, the schedule, race results, and other logistical items. The site will look like the team, featuring similar coloring and branding. It will also be cool and inviting to visitors, serving as a recruiting tool. All this will be accomplished by directing some members of the target audience to one part of the site and other members to another part. Thus users interested in the social or fun aspects of the team (Morgan and Amber) will be directed to pages that mostly feature pictures and race results. Those interest in logistics (Rachel and the coaches) will be directed to pages that contain the schedule, news items, and financial information.

The value proposition is clear in the first draft, but it is poorly stated. It also gives some clue as to what the design might look like, but leaves several unanswered questions. In fact, the part about the design is really little more than re-stated criteria rather than what the design is really about.

Value proposition

First we will refine the value proposition. Each numbered item represents a new refinement starting with the original value proposition:

1. The mountain bike team web site will be one place where all members of the team can find what they are looking for. It will contain pictures, the schedule, race results, and other logistical items. The site will look like the team, featuring similar coloring and branding. It will also be cool and inviting to visitors, serving as a recruiting tool.

We need to specifically call out why we are building the site and how we can measure success:

2. The site will contain all the relevant information about the team. The site will convey the sense of adventure and comradery felt by the participants.

This clearly states the value proposition, but does not "sell it." We need to encourage the reader of it to believe that this is a worthwhile endeavor.

3. Currently team information is communicated through a wide variety of avenues. The new team site will simplify communication, being one-stop-shopping for parents and athletes alike. The team members currently has difficult explaining why it is so much fun. The new team site will capture the athlete's passion and excitement for their sport.

We are almost there. The only remaining problems are that the value proposition is too lengthy and disjoint. Can we bring it all together into a single cohesive package?

4. The new team site captures the spirit and logistical details of the team into a single location providing one-stop-shopping for all of the team's communication needs.

The fourth attempt captures the essence of the value proposition without getting too bogged down in details and without skipping the core ideas.

Design approach

Next we will refine the design approach. As with the value proposition, each numbered item represents a new refinement of the original idea:

1. All this will be accomplished by directing some members of the target audience to one part of the site and other members to another part. Thus users interested in the social or fun aspects of the team (Morgan and Amber) will be directed to pages that mostly feature pictures and race results. Those interest in logistics (Rachel and the coaches) will be directed to pages that contain the schedule, news items, and financial information.

This has too many details and is also too difficult to follow. Is there some overarching design idea that we can focus on?

2. The site will direct social users to one set of pages and logistical uses to other pages.

This is certainly more focused, but misses some important details.

3. The site's top-level navigation will direct social users to media items and direct the logistical users to the schedule, news items, and financial information.

A few of these words are difficult to understand. We will bring our personas back into the process.

4. The site's top-level navigation will direct Morgan and Amber to pictures and race results while directing the Rachel and the coaches to logistical information (schedule, news items, and financial information).

Bring it all together

The new team site captures the spirit and logistical details of the team into a single location providing one-stop-shopping for all of the team's communication needs. This will be accomplished by the navigation structure which will direct Morgan and Amber to pictures and race results while directing the Rachel and the coaches to logistical information (schedule, news items, and financial information).

Comprehension check

Quiz 1 - 3 **Find three examples**

Find three examples of executive summaries, mission statements, or personal mission statements that meet the complete, concise, convincing, and repeatable requirements of an executive summary. For each one, provide the full text and cite where it came from.

Quiz 4 **Write an executive summary**

What is your personal mission statement, executive summary, or prime directive? Again, make sure it fits the complete, concise, convincing, and repeatable criteria.

Problems

Problem 1 **Weather web site**

You are tasked with creating a new web site to display weather information. You are given the following persona and scenario:

- Persona: Rachel is a stay-at-home mom who uses the internet every day for socializing, news, and shopping. Though there is always a computer on hand, she seldom can sit for more than five minutes before being interrupted.
- Scenario: It is Sunday night and Rachel is making a plan for her week. A big part of this plan is the weather, of course. By recommendation of a friend, she decides to try www.weatherman.com to plan her week.

Come up with a list of a dozen broad directions your design could go. List a half dozen value propositions and a half dozen big design decisions. These are the beginnings of an executive summary

Problem 2 **CB radio**

Imagine you were newly hired to a company that produces CB radios. Knowing your expertise developing user interfaces and feeling that the UI of their current CB radio is a bit dated, they task you with proposing a new design. Use the persona and scenario you developed in a previous lesson. Come up with a list of a dozen broad directions your design could go. List a half dozen value propositions and a half dozen big design decisions. These are the beginnings of an executive summary

Problem 3 **Stopwatch app**

Write a single executive summary for a stopwatch app. Make sure it meets the complete, concise, convincing, and repeatable criteria. Please use the persona and scenario developed in a previous lesson.

Problem 4 **Infotainment system**

Write a single executive summary for an all-"all glass" instrument cluster for a modern car. All the controls are reachable through a single touch interface. Make sure it meets the complete, concise, convincing, and repeatable criteria. Please use the persona and scenario developed in a previous lesson.

09 Design Process

There is nothing quite as intimidating as a blank sheet of paper for an author or a clean canvas for a painter. Where to start? What separates an "expert artist" from a novice is not necessarily his raw talent. Equally important are the tools at his disposal and the way in which he channels his creativity. Designing user interfaces is similar in many ways. While our job is clearly more technical than that of an artist, we still need to learn to channel our creativity and "fill our tool bag" in order to be truly effective.

Design as a Search Problem

For the sake of discussion, let's assume we have a single one-dimensional scale measuring the quality of a given design. Thus we can ask our oracle to evaluate a design and "poof!" we get a nice scalar value as a result. Now consider the scenario where we are trying to find the optimal font size for a web page. In other words, we only have the ability to manipulate a single variable. In this scenario, the design problem can be represented with a simple graph.

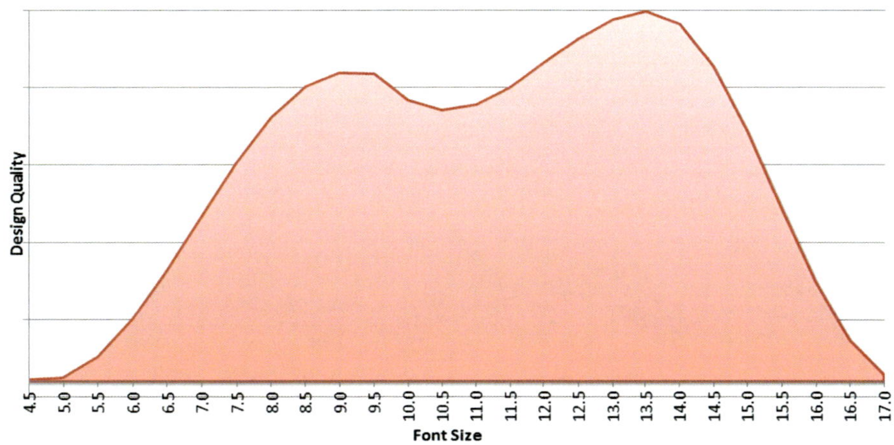

Figure 09.1 - Design search-space for a one-dimensional problem

Notice how very small and very large font sizes yield a poor design quality. In the middle are two local maximums (font size around 9.5 and another close to 13.5) with the larger being slightly better. The job of the designer is to find the best possible value.

In many ways, design is a search problem. Of course, the number of variables can be very large and their interactions can be quite complex. This means it is impractical or often impossible to search every possible value. Also, our oracle tends to be expensive and somewhat unreliable. Because there is a bit of human subjectivity built into all evaluation methods, we cannot be sure we will get the same value with two measurements or that the measurement we obtain is an actual reflection of the quality of the design. Nevertheless the role of the designer remains the same: find the best or optimal values for each design parameter.

At this point it is important to point out that we will never find the optimal design for a given problem. We can only hope to find a good design or, more commonly, a good enough design. There is always room for improvement. While this is certainly a humbling realization, it is also a source of great hope. We can always do better tomorrow!

Fitness landscape

As mentioned previously, most real-world design problems are characterized by a very large number of variables. It is typically unfeasible for all possible design alternatives to be explored. Instead, the designer is tasked with finding a "good" design that is "better than most." The set of all possible designs is called the fitness landscape. The **fitness landscape** is a multi-dimensional space (the exact number depending on the number of variables) where each point is characterized with a value representing the quality or fitness of the design at that point. While it is impossible to determine the "true" fitness value for a given point, we can get a good estimate through the evaluation process (heuristic, expert, or experimental).

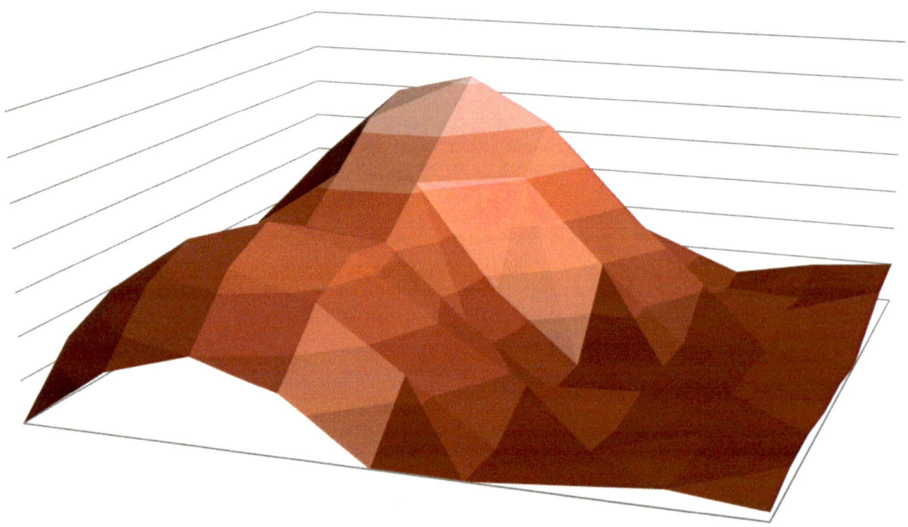

Figure 09.2 - Fitness Landscape where the higher the "hill," the better the design quality for those variable settings

There are several properties of a fitness landscape worth mentioning: smoothness, uniform slope, and multiple hills.

Smoothness The first property, smoothness, states the fitness level of neighboring variable values tends to be similar. In other words, changing one or two variables by a small amount is unlikely to have a dramatic influence over the overall quality of the design.

Uniform Slope The uniform slope property states that the influence of a small change in a variable tends to be replicated as the variable is changed further. This leads to the strategy of **hill climbing**: continue making changes in a given direction until the quality begins to fall off.

Multiple Hills The multiple hill property states that reaching a local maximum is no guarantee the best possible design will be found. To find that value, it is often necessary to find another hill to climb. Of course, there may be an extremely large number of hills in a given fitness landscape depending on the complexity of the problem being addressed.

The Iterative Design Problem

Given the complexity of the fitness landscape and the difficulty in ascertaining a single measurement, how can a designer be expected to find a good design? Drawing from our knowledge of search derived from our Computer Science background, a couple ideas should come to mind...

Depth-first search

The first search strategy is depth-first search. In the context of design, depth-first search can be defined as: for a given basic design, refine the variables until a local maximum is found. This process involves systematically identifying shortcomings in a design, removing them, and re-evaluating. As this process continues through many iterations, the evolving design slowly "walks" up a hill in the fitness landscape. To be effective in this process, the designer needs to be conscious of how each change influences the overall design quality.

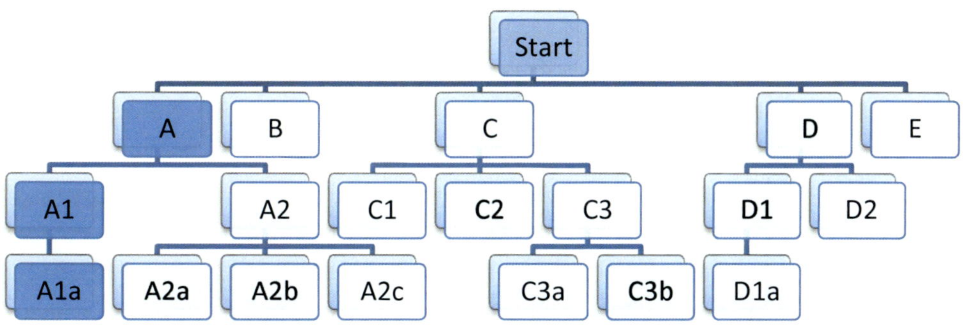

Figure 09.3 - Depth-first search through the design space. One idea is refined until it is perfected. Then the next idea is refined and so on. In this case, A, then A1, then A1a are explored in turn. After this line is exhausted, then A2 and A2a will be explored.

Use depth-first search when you believe you are on a good design track and want to refine your thinking. Here you take lessons learned from your previous design iteration and make a series of small, incremental design improvements. The key word for depth-first search is "refinement."

Depth-first search leverages the smoothness and uniform slope property of the typical fitness landscape. However, it does not address the multiple hills property.

Breadth-first search

The second strategy is similar to the classic breadth-first search. In this case, the designer tries out many wildly-different design alternatives. Each alternative is evaluated in hopes of chancing upon a good-sized hill in the Fitness Landscape. This process is commonly called **brainstorming**. In the breadth-first search mode, emphasis is placed on scoping out the big design decisions. To be effective in this process, the designer needs to think outside the box, use creativity, and momentarily forget technological or resource constraints. One common approach is to identify a design maximizing each successive variable of usability in turn. This yields eight designs with which to start the process.

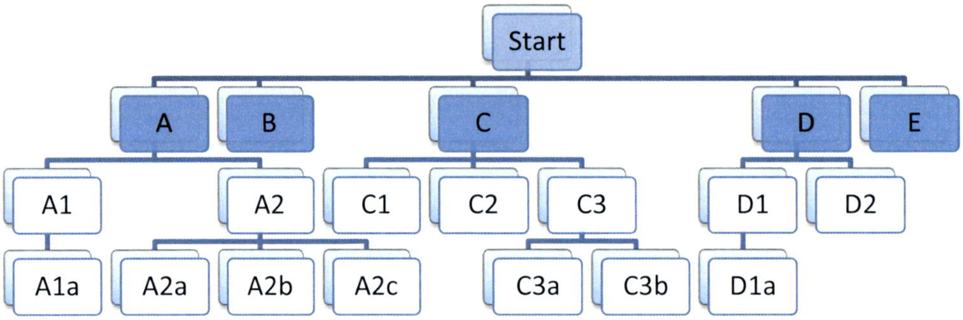

Figure 09.4 - Breadth-first search through the design space. The big ideas are explored first before any degree of refinement is done. Only after A, B, C, D, and E are explored will A1, A2, C1, etc. be explored.

Use breadth-first search when you are unsure where to go next with the design. Here you get many ideas out there, hoping to find a productive direction. Breadth-first search addresses the multiple hills property of the fitness landscape, but does not leverage the smoothness nor uniform slope property.

Hybrid approach

Since both the depth-first and the breadth-first approaches fail to address all the properties of the design search space, clearly another methodology is needed. Thus, most designers follow a hybrid approach. They first conduct a breadth-first survey of the landscape to try to get a "lay of the land." Once the most promising design alternatives are identified, they are refined using the depth-first approach. Note that this is seldom a two-step process; often the designer needs to go back to square one or, at a given juncture in the refinement process, conducts a miniature breadth-first investigation of the alternatives.

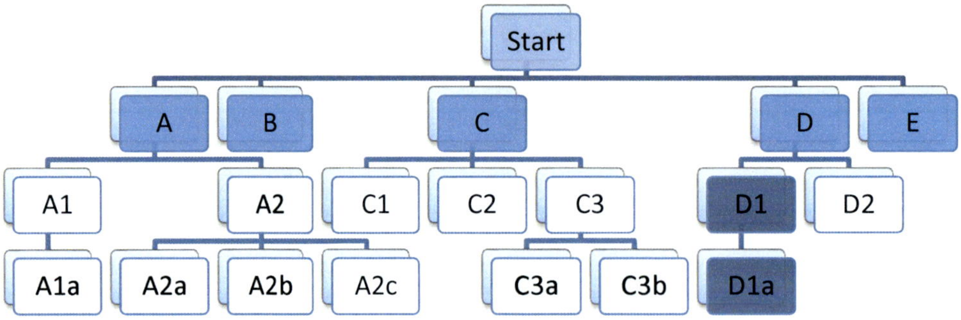

Figure 09.5 - The hybrid search approach involves first conducting a breadth-first search of the design space (A, B, C, D, E), then refining the best alternative (D1, D1a). Note that this is often a multi-step process. For example, C might be investigated next, necessitating another breadth-first search (C1, C2, C3).

It is important to note that the design process is not like a horse-race where design alternatives are competing for the honor of being chosen for implementation. Instead, the designer uses insight gained from his various trials to guide the process. If one promising lead turns into a dead end after a few iterations of refinement, then components of that design could still be reused in a completely different branch of the design tree.

Examples

To illustrate this process, two examples will be shown. The first example will be presented at a high level narrative describing the process with broad strokes. The second example will illustrate the drafting and evaluation process the designer goes through as he iterates through many design alternatives.

Example 1: Smart phone interface

Consider a designer tasked with identifying a pointing mechanism for a new smart phone. Rather than jumping to the design her competitors are using, she begins the process with a breadth-first search. The list of alternatives after the first round include a mouse, a pen, finger touch interface, cursor keys, track ball, and voice. After an informal heuristic evaluation of these alternatives, she settles on the touch interface and cursor keys.

Our designer next refines the cursor key alternative because it had a slightly better score than the touch interface. Again, unsure where to go, she does some more brainstorming and comes up with another set of options: arrow keys similar to a keyboard, a rocker switch similar to a game controller, and page buttons similar to an eBook reader. Serious problems with the first two rule them out, leaving us with the page buttons as the best alternative.

Deep Dive:

The process of taking a design all the way to the end of the refinement process to see how it plays out.

From here, our designer conducts a **deep dive**. A deep dive is the process of taking a design all the way to the end of the refinement process to see how it plays out. This process can be very educational to the designer, yielding insight into the problem that is difficult to ascertain from high-level sketches. From this process, she realizes the importance of hand position and the efficiency issues associated with reading large quantities of text. While she has learned much, she was unable to address a host of other important scenarios causing her to abandon this line of inquiry.

Our designer then returns to another of her high-level design alternative: the touch interface. Again, many design decisions are explored (point-centric or swipe-centric designs, for example), and one promising alternative is refined in a deep dive. While our designer becomes confident this design can address most of the key scenarios, the reading scenario is still lacking.

Recalling how the page button design did well in the reading scenario, and realizing the touch design did poorly, our designer makes an attempt to combine the best components of the two. Because they rely on different technology, this becomes an easy process. After a few more refinement iterations to make sure the features are fully integrated, she is ready to take a prototype to the usability lab.

Example 2: Team web site

Recall the web site project for the local high school mountain bike team? At this point in the design process, the most important tools are the criteria rubric and the executive summary. Nevertheless, we must always remember the persona and scenario. The rubric is the following:

	Exceptional 2	Good 1	Acceptable 0	Flawed -1	Horrible -2
Important Info 40%	Rachel is nearly certain to not miss important information	The important info is prominently displayed	It is clear where the important information is posted	One can find the important information with effort	It is very difficult to find important information
Motivating 30%	The site draws people in and encourages sharing	The site is pleasing to the eye and interesting	Everything is professional	The site is boring	Unprofessional, ugly, or punishing
Learnable 10%	Everyone figures it out immediately	Most users will figure out the site in just a minute or two	Familiar and learnable, but more difficult than necessary	Some users will never discover some features of the site	Many users never discover some features of the site
Representative 10%	The site captures the culture and feeling of the team	The site resonates with members of the team	The data in the site is accurate and trustworthy	Some aspects of the site are misleading or incorrect	The site misleads visitors as to what the team is all about
Understandable 10%	Extremely simple and straightforward	Most users would call the site "simple"	As one would expect	More complex than necessary	Requires much more work

The executive summary is:

The new team site captures the spirit and logistical details of the team into a single location providing one-stop-shopping for all of the team's communication needs. This will be accomplished by the navigation structure which will direct Morgan and Amber to pictures and race results while directing the Rachel and the coaches to logistical information (schedule, news items, and financial information).

The design process will begin by doing a breadth-first search through the design space by doing some brainstorming. Initially many different navigation ideas will be tried, each of which evaluated. Next one or two promising lines will be explored in detail. Finally, the design will be refined until all the ideas are fully flushed out.

Round 1: Long Scroll

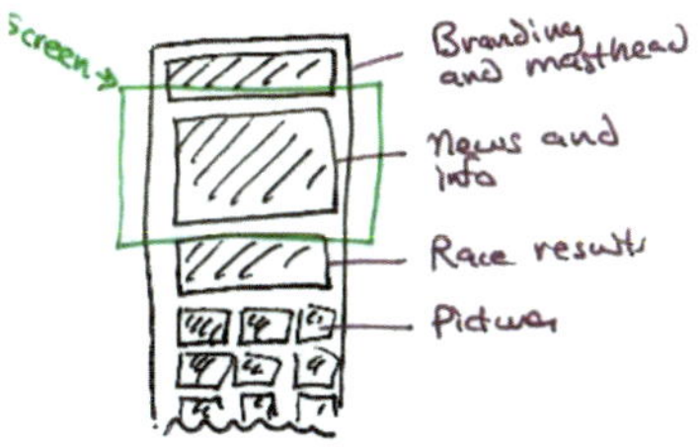

Figure 09.6 – Design idea consisting of a single long page

The first design alternative is to put all the data on a single page. Here the user only needs to scroll down to find everything on the web site. The green rectangle represents what can be seen on a single screen at a time. Our heuristic evaluation of this design yields:

Criteria	*Weight*	*Score*	*Comment*
Info	0.4	2	Info is at the top of the screen
Interesting	0.3	-1	Boring and static
Discover	0.1	1.5	Easy to figure out
Representative	0.1	0	Accurate
Understandable	0.1	0	Simple but overwhelming

The overall score for this design is 0.65.

Round 1: Sub-pages

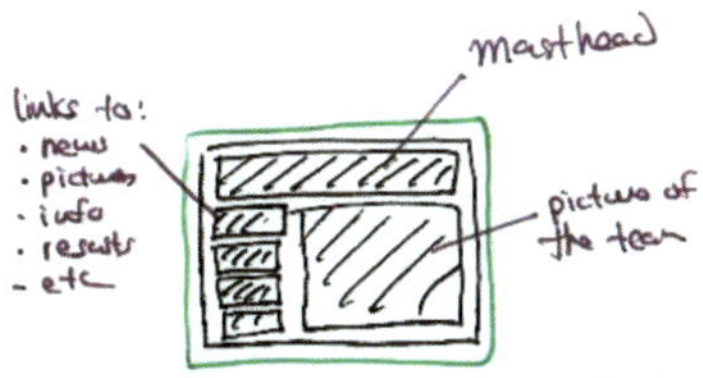

Figure 09.7 – Design idea ofmany small pages

Our next brainstorming idea is to have a collection of static pages, each one liked to from the home page:

Criteria	*Weight*	*Score*	*Comment*
Info	0.4	-1	Info is at the top of the screen
Interesting	0.3	0	Boring and static
Discover	0.1	1	Easy to figure out
Representative	0.1	0.5	Simple but overwhelming
Understandable	0.1	0.5	User might lose her place

The overall score for this design is -0.2

Round 1: Separate by Persona

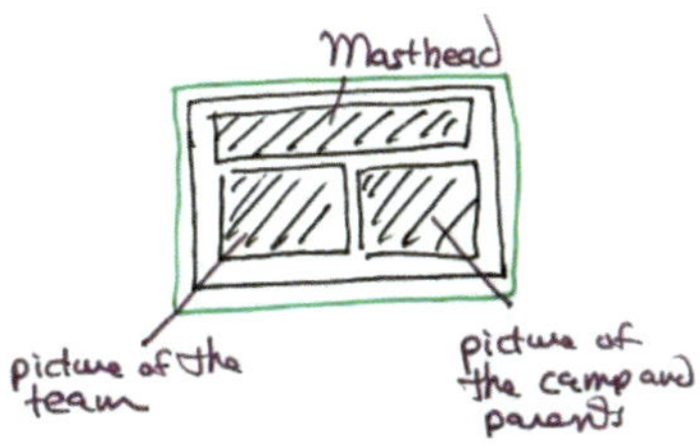

Figure 09.8 - An attempt to segment the target audience

Another brainstorming idea is to try to direct the different personas to different places of the site. Here we will present information relevant only to one persona at a time.

Criteria	*Weight*	*Score*	*Comment*
Info	0.4	-1	Might miss the news
Interesting	0.3	-0.5	Boring and static
Discover	0.1	-1	Many might miss the site's meaning
Representative	0.1	0.75	Resonates a bit…
Understandable	0.1	-0.75	Novel, but some might get lost

The overall score for this design is -0.5

Round 1: Auto-scroll

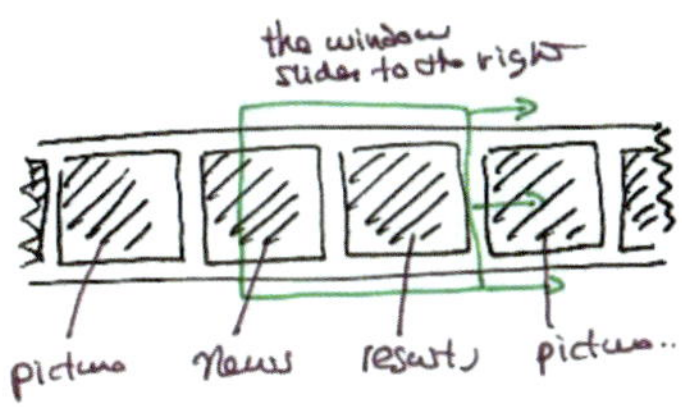

Figure 09.9 - One long horizontal scroll

This is a variation of the "All in one" idea, but here we will auto-scroll through the various pieces of information. This includes pictures, news, race results, etc.

Criteria	*Weight*	*Score*	*Comment*
Info	0.4	-1.5	Takes much waiting to see anything
Interesting	0.3	1	Entertaining
Discover	0.1	0	More difficult than necessary
Representative	0.1	1	Resonates with the pictures
Understandable	0.1	-1.5	Weird, novel, and no sense of position

The overall score for this design is -0.35

After the initial round of brainstorming, it is time to take the best ideas and perform some refinement. Variations of the single-page design did not score very well. We will next focus on the first-level navigation

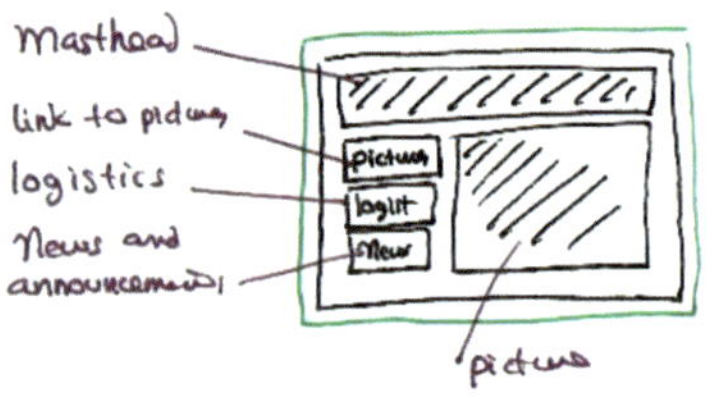

Figure 09.10 - Refinement of the nav-bar

Navigation Refinement: Vertical Nav-bar

The idea here is to have a collection of menu items along the left edge of a static page linking to specific information. We will also have a large and prominent picture dominating the center of the page. The top-level navigation will read: Pictures, Logistics, and News.

Criteria	*Weight*	*Score*	*Comment*
Info	0.4	-1	Might miss the news
Interesting	0.3	0	Professional but boring
Discover	0.1	1.5	Hierarchy is better
Representative	0.1	0.5	Picture could be improved
Understandable	0.1	0.75	Simple, never get lost

The overall score for this design is -0.125.

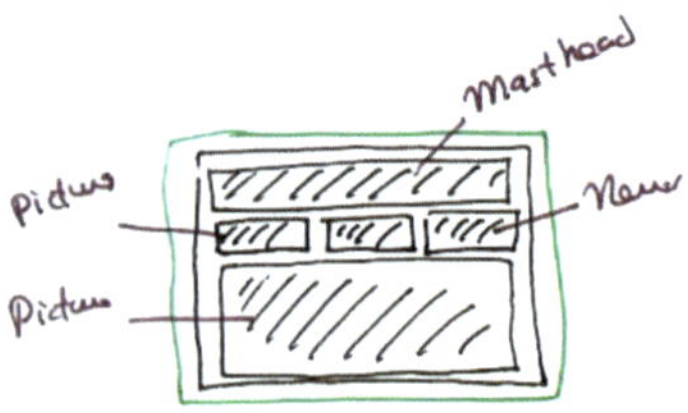

Figure 09.11 - Another nav-bar idea

Navigation Refinement: Horizontal Nav-Bar

This is a variation of the vertical nav-bar and has many of the same issues. However, we notice immediately that we have much more space for the picture. Not only that, the picture itself is not "boxed in" by the nav-bar. This gives us an opportunity to do something new here.

Criteria	*Weight*	*Score*	*Comment*
Info	0.4	-1	News might be missed
Interesting	0.3	0.5	More space for the picture!
Discover	0.1	1.5	Same hierarchy as before
Representative	0.1	1	More space for the picture!
Understandable	0.1	0.75	Simple, none will get lost

The overall score for this design is 0.075

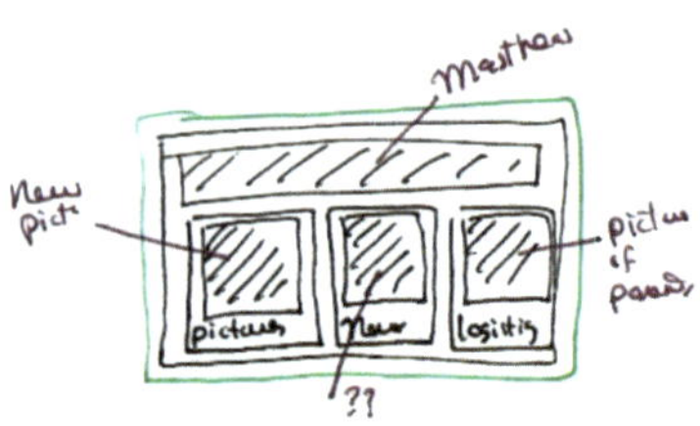

Figure 09.12 - Navigation performed by gallary rather than by nav-bar

Navigation Refinement: Gallery Navigation

Here we make navigation much more prominent, using not only a label but also a large picture. It makes it very easy to select a sub-page, but also removes our large picture.

Criteria	*Weight*	*Score*	*Comment*
Info	0.4	1	News is more prominent
Interesting	0.3	-1	Loss of picture is boring
Discover	0.1	1.5	Same hierarchy
Representative	0.1	-1	Needs a picture!
Understandable	0.1	-.25	Novel, none will get lost

The overall score for this design is 0.125

We have solved some problems, but have yet to bring all these ideas together. Somehow we need to make the site interesting and also easy to get the necessary information. Clearly we need another round of design.

Figure 09.13 - Refinement of how the pictures are presented

Picture Refinement: Manual Scroll

Going off of our horizontal nav-bar design earlier, we will add some interesting by hosting more than one picture on the home page. In fact, we will allow the use to scroll through a collection of hand-picked pictures representing different aspects of the team and the members.

Criteria	*Weight*	*Score*	*Comment*
Info	0.4	-1	Might miss the news
Interesting	0.3	1	More interesting from more pictures
Discover	0.1	1.5	Good hierarchy
Representative	0.1	1.5	The pictures capture the fell of the team
Understandable	0.1	1	Simple navigation, never get lost

The overall score for this design is 0.3.

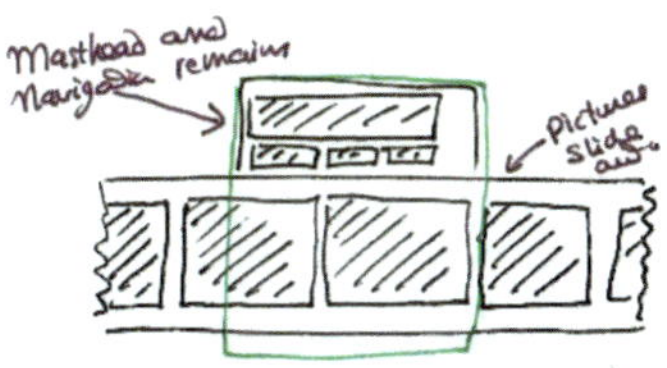

Figure 09.14 - Utilizing the auto-scroll feature

Picture Refinement: Auto-scroll

This is a variation of the manual-scroll design above. The only difference is that we will auto-scroll through the pictures with a 2-3 second pause between each viewing. This will be dynamic and more interesting

Criteria	*Weight*	*Score*	*Comment*
Info	0.4	-1	Might miss the news
Interesting	0.3	2	Auto-scroll is fun and interesting
Discover	0.1	1.5	Good hierarchy
Representative	0.1	2	Everyone will see all the pictures
Understandable	0.1	0.75	Simple, none will get lost

The overall score for this design is 0.625

The design is progressing well. We have the top-level navigation figured out, we have found a good way to represent feel of the team, and we have made the home page interesting. The only problem remaining is how to get the urgent information to the users. Recall our critical scenario: Rachel checking the site just before heading off to a race. If it is not apparent that the race got moved, she will have a very bad experience.

Figure 09.15 - First attempt at a notification feature

Notification Integration: Shadow-box

Here we will throw up a shadow-box on the rare occasion that something critical has come up. It will therefore be impossible to miss!

Criteria	*Weight*	*Score*	*Comment*
Info	0.4	2	Impossible to miss
Interesting	0.3	-1	Distracting
Discover	0.1	1.5	Good hierarchy
Representative	0.1	-1	Alert has too much precedence
Understandable	0.1	1	Simple but confusing

The overall score for this design is 0.65. This is our best score yet, but there are so many negative consequences. Can we have the best of both worlds?

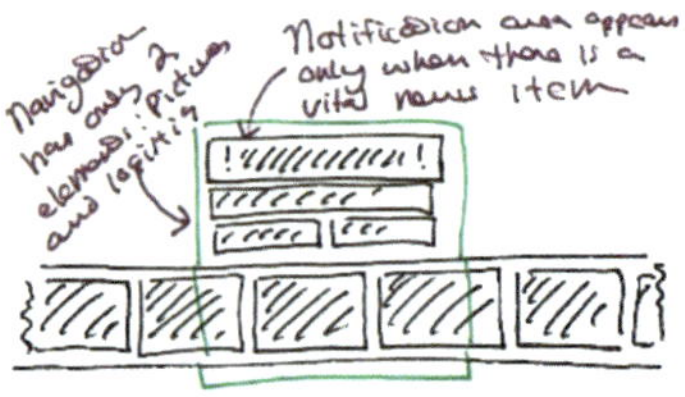

Figure 09.16 - Close to the final design

Notification Integration: Notification Bar

Rather than covering content, we will just insert a block of text above the masthead if there is an urgent notification. This will be prominent but will not displace any content.

Criteria	*Weight*	*Score*	*Comment*
Info	0.4	1.5	Very hard to miss
Interesting	0.3	2	Love that auto-scroll
Discover	0.1	2	Navigation is easier with 2 items
Representative	0.1	2	Representative pictures
Understandable	0.1	1.5	Simplest design yet

The overall score for this design is 1.75. Here we truly got the best of both worlds. All of our Priority 1 criteria are met and most of our Priority 2 as well.

The design process for this site is not complete. We still need to go through the individual pages and sub-pages. However the same process is followed. When finished, we have the following page:

Figure 09.17 - Mockup of the final design

Comprehension check

Quiz 1 **Identify the best design approach**
At the beginning of the design process, the designer is attempting to get a feel for the Fitness Landscape. Nothing is yet set in stone; he wants to know which direction to channel his efforts.

Quiz 2 **Identify the best design approach**
The design is nearing completion and some refinement is in order.

Quiz 3 **Identify the best design approach**
I am trying to map out a design strategy for an entirely new feature. This strategy will take weeks to complete and include many iterations.

Quiz 4-6 **Fitness landscape**
In your own words, define each of the following:

- The Smoothness property
- The Uniform Slope property
- The Multiple Hills property

Problems

For each of the two problems, make sure you can demonstrate:

- **Breadth-first search**: Have examples of wildly different designs exploring the full gamut of possibilities.
- **Depth-first search**: A few of the better designs should be refined to work out problems
- **Evaluation**: Each iteration of the design process should be evaluated according to the criteria previously identified. Score the designs and comment on their strengths and weaknesses.
- **Prototype**: The final prototype should be detailed enough to clearly illustrate the key design characteristics.

Problem 1 **Stopwatch app**
Demonstrate the design process by designing a stopwatch app. Please use the persona, scenario, evaluation criteria, and executive summary developed from previous problems.

Problem 2 **Index Feature**
Demonstrate the design process by designing the index feature on an electronic textbook. Please use the persona, scenario, evaluation criteria, and executive summary developed from previous problems.

10 Prototypes

There exists a conundrum in the design process: you can't evaluate a design until it is built, and once something is built it is often difficult to make changes to the design. As a result, interface designers are often asked to design a feature without any real idea of how the user will react to it. Large amounts of resources are then devoted to building this untried design. When it is finished, it is often too late in the development lifecycle to make needed changes and the development team has become invested in the existing design. Clearly a tool is needed to tackle this problem.

A prototype is a tool used to draft out a design for the purpose of working out details, evaluating quality, and communicating ideas with the product team. In most situations, the prototype for a feature or product is the focus of the design effort, representing all the ideas the team is considering for the final product.

Fidelity

The purpose of the prototype is to be a reflection of the current thinking of the design team. This means it needs to be both as accurate as possible, neither lacking important details nor containing details that have yet to be decided. Perhaps it is best to illustrate this with two examples.

Imagine a design team working on a new mobile banking app. It is early in the design phase so few big decisions have been made. Nevertheless, the stakeholders want a report. To satisfy management, a quick prototype is built. The team lead explains to the graphic designer what has been decided and asks him to come up with a few sketches. The graphic designer compiles and produces a half-dozen drawings incorporating the team lead's ideas but also quite a few of his own. Just before the meeting, the graphic designer shares his drawings with the team lead and the lead is delighted: the drawings look great! In fact, everyone in the stakeholder's meeting is impressed and ends the meeting with "great, build it!" As the lead leaves the meeting, he realizes his mistake. Those drawings, containing many design features that have not been vetted, have become the final design in the eyes of the stakeholders. In other words, the prototype was not an accurate representation of the design team's thinking. It was, in fact, misleading.

Learning from this mistake and working on a different product some time later, the same team lead has been asked to present a report to the stake holders. This time, the product is nearing completion and only a few small details have yet to be worked out. Determined to not make the same mistake, he brings only rough sketches of the design. He has a working prototype on his tablet, but does not want to share it with the stakeholders least they get overly invested in the unfinished product. As he presents the rough sketches to the stakeholders, they are not impressed. They thought the team was further along in the development process. As they ask questions, the team-lead pulls out a pen and updates his sketch to reflect decisions that have been long since made. To the stakeholders, it looks like he is winging it.

Fidelity:

The degree of detail present in a draft of a design.

In both these examples, the fidelity of the prototype failed to reflect the firmness of the decisions by the design team. Prototype fidelity is the amount of detail contained in a design. A high-fidelity prototype is nearly pixel perfect, reflecting the look and behavior of the final product. A low-fidelity prototype is a rough sketch, with dimensions, control placement, color, and other details clearly not specified. As a general rule, the fidelity of a prototype should increase as the project progresses. Initially it should be very rough and low-fidelity. Near the end, it should contain screen-shots from the actual product.

Types of Prototypes

There are many prototyping tools and techniques employed with success by design teams around the world. In many ways, the type of prototype is less important; it only matters that it successfully communicates your thinking. That being said, the more tools and techniques you have, the more informed you can be when it comes time to prototype a design of your own. To this end, we will discuss three of the most common prototyping techniques (storyboard, paper prototype, and interactive mockup) and then briefly introduce a few others.

Storyboard

A storyboard is a collection of drawn scenes representing a scenario. Originally developed in the movie industry to capture the plot and scenes of a movie, it has been adapted by user experience designers for the same purpose. There are several essential characteristics of a storyboard:

Scenes	A collection of moments or states in the scenario representing the most important interactions of the user with the interface
Drawings	A depiction of a scene coupled with the actions of the user interacting with it
Narrative	A textual description of what is going on during a scene
Transitions	What is happening between the scenes? This needs to be implied by the drawings (a click of the mouse for example) or stated in the narrative
Linear	Only a single instance of a scenario is explored in a storyboard. There are no variations

The following is an example of a storyboard for a simple interaction with the Windows Media Player. The user will boot the player, select a playlist, play the song, and then minimize it as she continues to work.

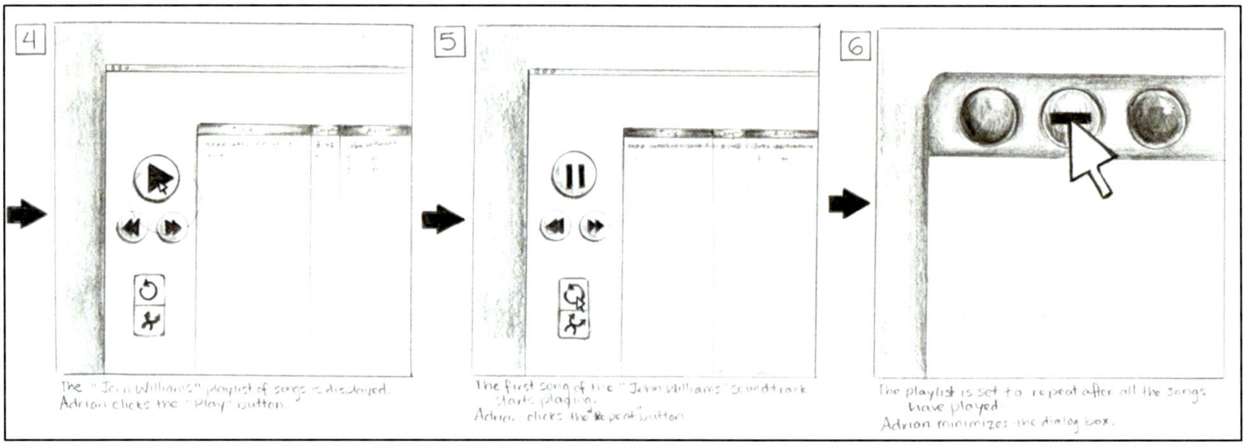

Figure 10.1 - Storyboard in pencil and paper

Note how the storyboard closely follows the scenario; both are trying to represent exactly the same thing. As an aside, storyboards are now commonly done in a slide deck, enabling edits to be made easily and for easier sharing to large groups of people. The same can be done digitally using PowerPoint, web pages, or a similar technology:

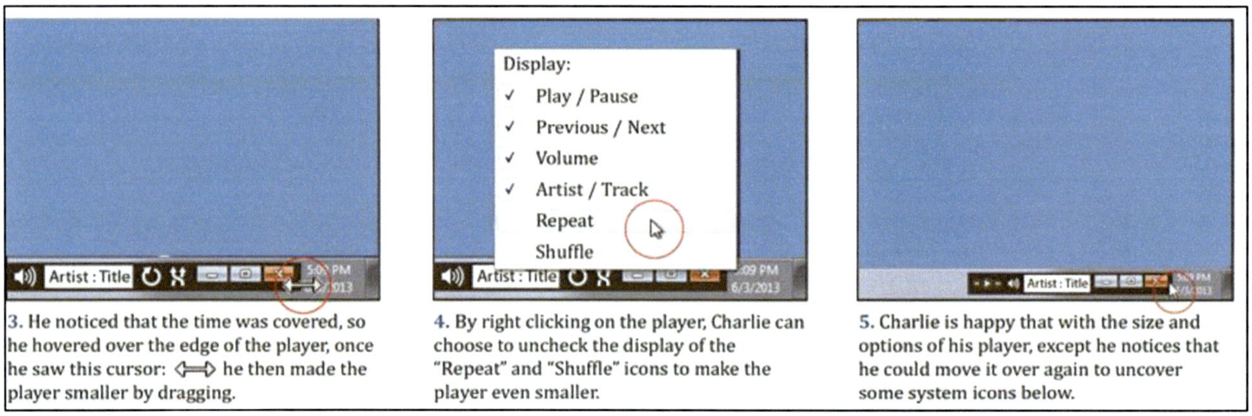

Figure 10.2 - Storyboard in PowerPoint

Paper Prototype

A paper prototype is a mockup of an interface complete with semi-interactive elements created out of paper or a similar convenient medium. Usually paper prototypes are made with a single drawing or image representing each screen or window. User data is represented with Post-It notes stuck on top of the screen or window drawing. Additionally, interactive elements such as listboxes and scrolling text are added by placing them on top of the main drawing. There is no limit to the types of interactions that can be captured by a paper prototype; just use your imagination and experiment.

The essential characteristics of a paper prototype are:

Screens or Windows A representation of each screen, window, or dialog that the user is likely to encounter in a given interaction. Each should be drawn on a separate piece of paper

Interactive Elements Small cutouts of paper representing the different states of controls. It is usually helpful to have a few blank cutouts so user data can be drawn and added to the interface

Manipulator A knowledgeable person who runs the prototype. If, for example, the user clicks on a button (by tapping the drawing with his pen), the manipulator would grab the drawing representing the dialog that would come up and places it on top of the currently displayed drawing

Nonlinear Many alternatives can be explored in a paper prototype. The path followed through the interface is up to the user

The following is an example of a paper prototype:

Figure 10.3 - User interacting with a paper prototype

Note that, unlike a storyboard, a paper prototype involves the active participation of the user. It is possible to conduct usability tests with paper prototypes and collect valuable data about designs that are not yet implemented. Also, many different use-cases can be explored in a single paper prototype. If the manipulator is quick about sketching new scenes or reflecting user data, the user can explore any part of the design in any order with a paper prototype.

Interactive Mockup

An interactive mockup is an electronic prototype built with a technology like presentation software, animation software, or HTML containing interactive elements. If there are enough interactive elements and the fidelity of the graphics is high enough, the prototype can be convincing to the user. The essential characteristics of an interactive mockup are:

Digital Renditions of the screens, windows, and interactive elements are digital. These could be high-fidelity drawings capturing the look of the product, or they could be hand-drawn sketches scanned for the purpose

Interactive Elements Hit-regions triggering actions are provided to give the user the perception of interaction. Typically these are done with hyperlinks leading to different scenes or slides

Note that interactive mockups could be linear or non-linear. An example of a linear interactive mockup is a simple slide deck where the user essentially clicks through a series of interactions. Unlike a storyboard, there is the perception of interaction as a representation of the mouse moves across the screen opening windows and manipulating controls. An example of a non-linear interactive mockup is an HTML prototype working much like the finished product.

Creating an interactive HTML prototype requires familiarization with the technology that is beyond the scope of this text. Creating an interactive presentation software, however, is readily learned. Start with images of all the screens and interactive elements that are to be portrayed. Include a drawing of the mouse cursor. It may seem that getting these graphics is difficult. However, most designers spend a few hours and make a collection of common interactive elements such as buttons, list boxes, dialog boxes, and such. It is not difficult to make a collection of a few hundred of these, representing most of the interactive elements needed in the vast majority of all prototypes. In other words, it is not difficult to get these elements. Here, we will start by collecting images of a simple dialog we would like to demonstrate.

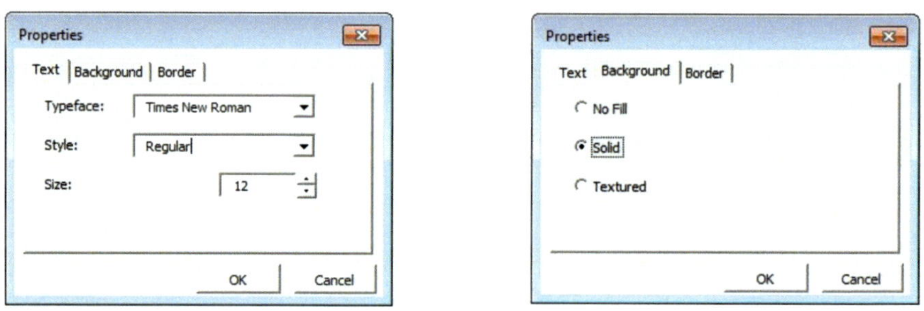

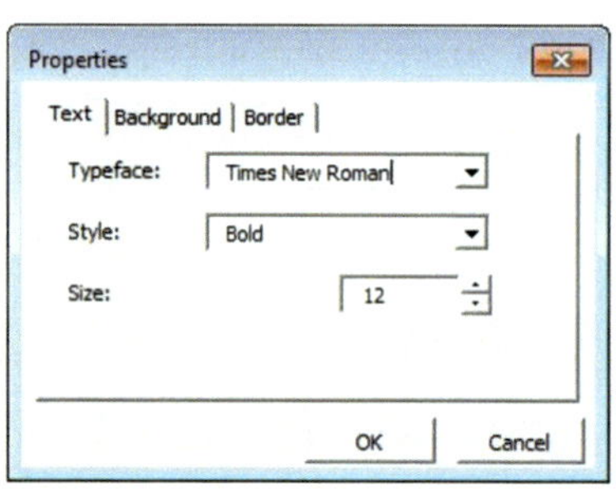

Figure 10.4 - A collection of elements used to create an interactive mockup

Next, place one of the screens on a blank slide with a plain white background. Add a mouse cursor image to the slide. Bring up the "Custom Animation" task-pane or palette. Select the mouse cursor image and select a "Custom Motion Path" to move the cursor image from its current location to the location of the interactive element to be clicked.

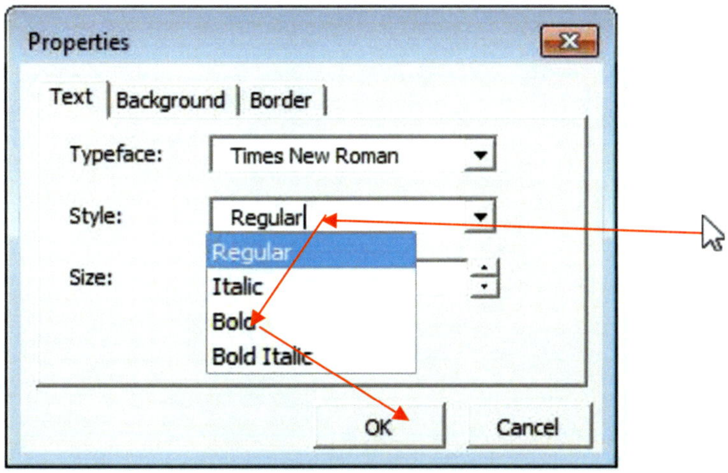

Figure 10.5 - A custom motion path animating the cursor across a picture of a dialog

Next, add an image representing what would happen if the element were to be clicked. In this case, a dialog will be opened. Again, from the Custom Animation pane, select "Entrance Effect : Appear." Test the animation to make sure that the perception of the user opening the dialog is complete. It may take a few tries to get the path and the timing of the mouse cursor animation right so it looks realistic.

Others

There are many, many different prototyping methods. While the most common are the storyboard, paper prototype, and interactive mockup, the following are also used occasionally:

- **Partial working system**: Using the same technology used to make the final product, the developers gets enough of the interface working to give the user the perception of a working system.
- **Wizard of Oz**: The user is presented with a rendition of the interface and "clicks" on interactive elements with her finger. An unseen Manipulator then alters the rendition (often by presenting a new paper or by altering the screen seen by the user) to give the perception of responsiveness.
- **Video prototype**: A non-interactive movie is created representing a user's interaction with the system. This behaves much like a storyboard.

Creating a prototype

Creating a prototype, be that a storyboard, a paper prototype, an interactive mockup, or any other format, falls in the same five steps.

1. **Scenario**: Everything starts with the scenario. For a storyboard or a video prototype, the scenario represents the script to be followed. For non-linear prototypes such as paper prototypes or partial working systems, the scenario will describe which interactive elements will be needed.

2. **Scenes**: What screens and data will be depicted in the prototype? Again, the scenario is the primary source to find these answers. However, as the prototype is flushed out, additional scenes are inevitably required. Each scene will require additional artwork.

3. **Interactive**: What elements will the user be interacting with? Each interactive element or control as well as the data the control represents will need to be depicted in the prototype. As with the scene identification, the scenario is the primary source but it is usually not detailed enough to capture all the interactive elements.

4. **Trial**: Perform a trial walkthrough. Does it all fit together or are there holes? Does the design make sense or are there problems already apparent. This is an important step in the prototyping process because it helps the designer identify problems or details which have not been fully flushed out.

5. **Details**: It is rare that the first draft of a prototype is good enough to fulfill all its needs. Make another pass or two to fill in details, smooth out rough edges, and refine the drawings. Remember, the more refined the thinking of the designer, the higher fidelity the prototype should be.

Examples

To illustrate the prototyping process, consider the high school mountain bike team site we have been working on the past few chapters. As you may recall, the scenario is:

> Rachel is preparing dinner when her children start coming home from school. While eating a muffin, Morgan asks her about the upcoming weekend. She wants to know who is driving and what they will be eating. Rachel walks over to her laptop on the dining room table and clicks on the Mountain Bike Team web site bookmark. Seeing the top-level navigation control, she clicks on the Schedule tab and quickly finds the next race. As she mentions a few details to Morgan, she looks for any changes or notes from the coach. There are none!
>
> As she talks with Morgan about the carpool plan (Rachel is driving this weekend) and the food (pasta again!), Rachel idly navigates to the Picture Exchange tab. She notices that some new pictures were added to the top of the list. As they firm up their meal plan, Rachel periodically calls Morgan over to look through some of the pictures. A few other curious family members peer over their shoulders as they laugh. Perhaps Morgan's brother will ride next year after all!

The goal is to present a design to the stake holders that they can understand, give meaningful feedback, and eventually approve. In other words, they do not want to get bogged down in the details of the design; they just want to feel that the design is on track and meets their needs.

This example will demonstrate three prototype techniques: a storyboard, a paper prototype, and an interactive mockup. First, a storyboard will be presented representing the "golden path" through the design which we hope the users will follow. From this initial prototype, we hope to understand if our understanding of the requirements is accurate and if our initial design ideas are on track. We fully expect to need to make big changes to the design at this phase.

Next, a paper prototype will be generated. Here we are quite interested in how the users will explore the interface. Will they seek functionality that does not exist? Will they be able to satisfy their needs? Will they be lost? Frustrated? Perhaps pleased? We will bring these findings to the stake holders with the hopes of getting preliminary approval of the design.

Finally, a high-fidelity interactive mockup will be created showcasing such details as font choice, use of colors, size & positions of the controls, and selection of pictures. These will be presented before all of the stake holders. The intent here is to get final approval of the design or to make any last-minute course corrections.

Storyboard

From the above scenario, we can identify several scenes:

1. Opens the web site and lands on the Home page
2. Notices that there are no notifications, navigates to the Logistics page
3. Reviews the time, location, and other details for the upcoming race
4. Navigates back to Home
5. Goes to the Pictures, then the race pictures from last week
6. Scrolls through the pictures

Since all the interactive elements in this design are just tap-able hit regions, no special interactions need to be represented. This means we can move straight to building the storyboard.

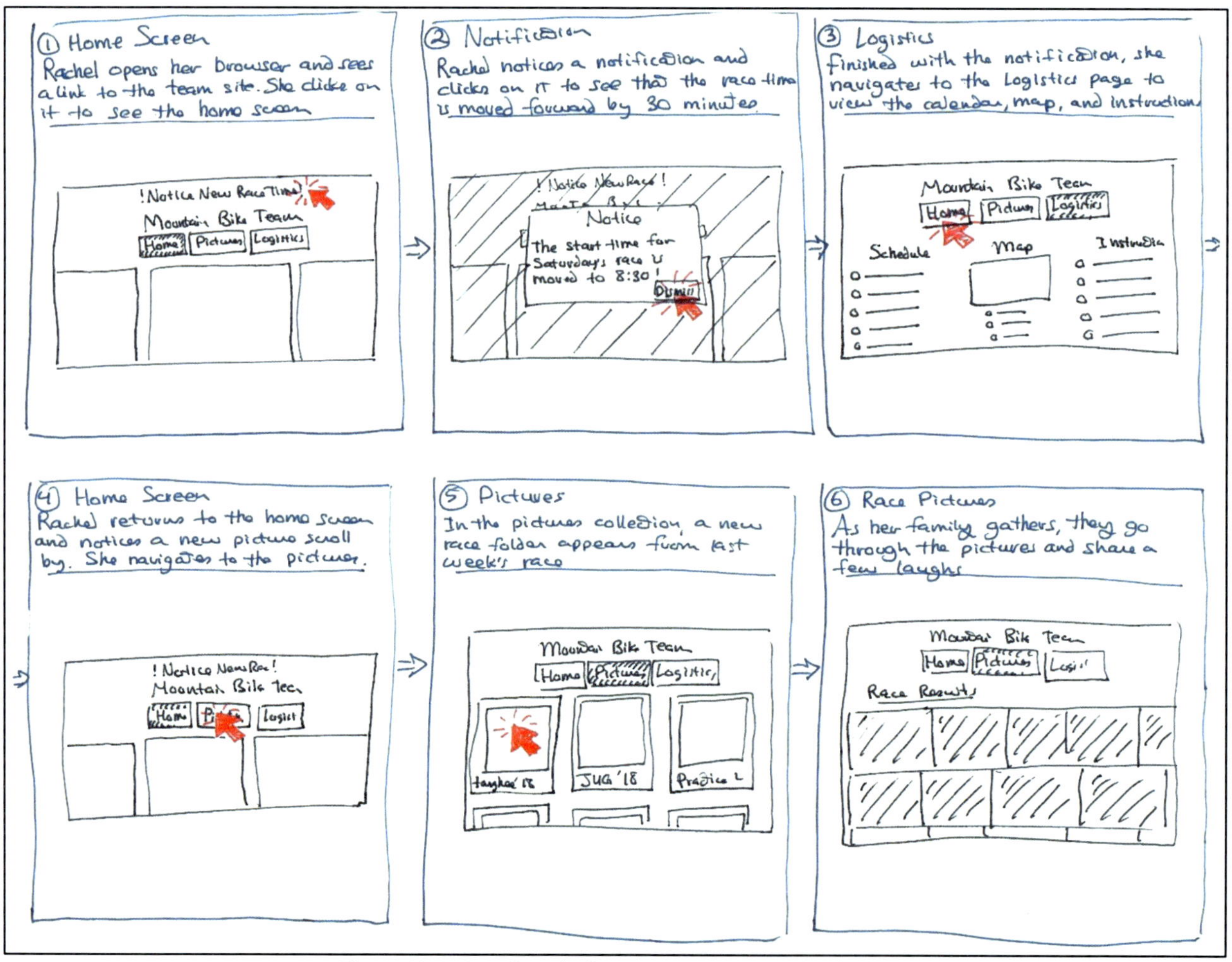

Figure 10.6 - Storyboard of a user interacting with a high school mountain bike team web site

Note that this is a low-fidelity prototype. The size of the text is not specified, the way rectangles and click regions are drawn is not specified, and all of the icons are very rough. However, the flow of control and the major concepts are fairly obvious.

Paper Prototype

To create a paper prototype for a project of this size, it is important to limit the scope. Creating a screen for every possible page containing all of the pictures and race information would be far too expensive. Instead, we will limit ourselves to just a few items.

1. **Scenario**: We will use the same scenario as the storyboard. The major exception is that Rachel will be using a tablet to view the web site rather than a laptop.
2. **Scenes**: Home page, notification, logistics page, pictures page, and the pictures from a specific race.
3. **Interactive**: We will simulate scrolling by creating pages which are larger than the device itself.

This will be a higher-fidelity prototype than the storyboard so we will create the graphics digitally. First, the iPad frame will be obtained from Apple with the image scaled to the correct size for an iPad screen. Next, the representative screens will be drawn in Adobe Photoshop at the screen resolution (1536 x 2048 at 264 dpi). The pieces will be printed and cut-out for the paper prototype.

Figure 10.7 - User interacting with a paper prototype

Interactive Mockup

The final prototype will be an interactive mockup created in presentation software. This will fortunately be easy because digital representations of the design were already created in the paper prototype. We will go through all 5 steps:

1. **Scenario**: We will use the same scenario as before: checking the schedule and them moving on to the picture exchange.
2. **Scenes**: The scenes will be the home page, the schedule page, and the picture exchange page.
3. **Interactive**: The interactive elements include the links on the home page to the schedule and picture pages, as well as the links back to the home page. We might do picture scrolling as well.
4. **Trial**: A collection of representative images were gathered. Digital mockups were created for each of the three scenes. These were all then put into a presentation software package to be linked with hyperlinks and animations. When finished, the designer went through the slide deck. Then a few co-workers viewed the slide deck. Finally, at the next team meeting, the slide deck was presented to the stake holders. They provided feedback and suggestions.
5. **Details**: After stake holder feedback was incorporated, then the prototype was re-created and presented to the stake holders for final approval.

Figure 10.8 - Interactive mockup of a high school mountain bike team web site

Comprehension check

Quiz 1 **Purpose of the prototype**

List all the reasons why one would want to use a prototype in the design phase of a project

Quiz 2-4 **Material**

List all the material necessary to create the following prototypes:

- Storyboard
- Paper Prototype
- Interactive Mockup

Quiz 5-7 **Advantages**

List the advantages and disadvantages for using each of the following prototype techniques:

- Storyboard
- Paper Prototype
- Interactive Mockup

Problems

Problem 1 **Storyboard**

Create a storyboard of the stopwatch app you designed in the previous lesson.

1. Include about a half-dozen scenes. The scenes should be derived from the scenario
2. In each scene, create a high-fidelity drawing of the design for the state represented in the scenes
3. In each scene, describe what type of user input occurs and how the system responds

Problem 2 **Paper Prototype**

Create a Paper Prototype for the stopwatch app.

1. Carefully draw then cut-out the mobile device.
2. Carefully draw the main screen of the stopwatch app
3. As appropriate, make the various controls interactive
4. Make the status indicators dynamic as well

Problem 3 **Interactive Mockup**

Create an Interactive Mockup for the stopwatch app.

1. Create a digital mockup of the design using multiple layers in Adobe Photoshop (or a similar program).
2. Save multiple images (probably JPGs or PNGs) representing the various scenes or elements.
3. Create a blank slide deck. Import the images from step 2.
4. Using animations, make the UI elements appear interactive

11 Specification

The development team working on a user interface problem consists of a variety of people working in different roles: the designer, the tester, the developer, and others. It is easy to think of them working together on a problem with adjacent offices and meeting daily to discuss the project. Unfortunately, this is almost never the case. Instead the members of a team are frequently separated by significant distances. More commonly, they are separated by time. It is often the case that the designer has finished working out all the details of the user interface long before the developer and tester have even been assigned to the project! How does a development team separated by space and time communicate under such circumstances?

Specification:

A document serving as the primary communication vehicle between members of the design team

The functional specification (otherwise known as the "spec") is the primary vehicle of communication between members of the development team. Its job is to answer what questions (e.g. "what will the team be building?"), why questions (e.g. "why is this product necessary and why this feature will satisfy the user's needs?"), and how questions (e.g. "how exactly are we going to build this product?"). Note that if the spec fails to communicate any of this information, the product will likely fail. This is true regardless of the quality of the ideas of the team members: if the ideas are not accurately communicated, they are wasted!

Usability Principles & Documents

A document, especially one as complex as a spec, has a user interface. While a traditional user interface is designed to be a communication vehicle between the system and the user, a document needs to be a communication vehicle between the ideas of the author and the reader. They are essentially the same problem.

Variables of Usability

The first step in seeing a document as a usability problem is to understand how the variables of usability relate to reading.

Efficiency How much effort is required for the user to receive the needed information from the document? Usual barriers to efficiency include the length of time required to read the document and how much mental processing is required to understand the presented concepts. Short, straight-forward, and direct documents are usually more efficient than those written with verbose, wordy, complex, and subtle writing styles.

Learnability How long will it take for the reader to figure out how the paper is organized and find what is needed? Readers presented with documents organized in a novel way need to figure out what all the components mean and how they relate to each other. Do the various document features encourage the reader to make those connections?

Familiarity Does this document resemble other documents serving similar purposes? Recipes, for example, typically have the ingredient list presented in a standard format on top and a numbered list of instructions on the bottom. A recipe following this convention does well with familiarity.

Simplicity Is the structure or the language more complex than necessary to communicate what the author wishes to say? While authors and especially poets often use complex language to make their text more interesting to read, this often has negative effects on the overall readability of the document.

Mapping Does the structure of the document encourage the reader to form a mental model of the subject like that which the author intends? Recall that the purpose of a document is to communicate the author's ideas into the reader's mind. The structure of the document, not just the words, should facilitate this transfer.

Motivation Is the document a pleasure to read? Some documents, like magazine articles and advertisements, draw the reader in. Others like a technical paper require effort.

Trust Does not apply in this context. The document does not hold the user's data and, in itself, is inert. This is not necessarily true with Tom Riddle's diary.

Visibility Some papers and books require the reader to constantly flip between pages to make connections and understand concepts. Others carefully lay out all the relevant information for a given topic in one location. Ideally a document should display all the information necessary to understand a topic on one page.

Just as the variables can be applied to the reading context, so can the rest of the Analysis and Design tools discussed in earlier chapters be applied to reading.

Analysis

Document analysis tools such as the target audience, task analysis, criteria identification, the iterative design process, and prototypes are useful in analyzing the usability of documents and as the first step in the design process.

1. **Target Audience**: Who is the reader for the document? What is her reading and education level? What are her expectations? How much time and patience does she have?

2. **Task Analysis**: What is the reader hoping to glean from the document? There has to be a purpose for her to read it. Without knowing this, we will have a difficult time meeting her needs.

3. **Criteria Identification**: By what metric will the reader judge this document? What makes one version of the document better or more effective than another?

Design

We can now begin the design process based on what we learned about the reader, the purpose for the reader viewing the document, and the metric by which we evaluate a document. This can be accomplished by iterating through several designs, making a prototype or a draft of the document, and finally writing the document. In other words, we will follow the same process for designing a document as we do for an interface.

4. **Designs**: Consider several layout options as well as ways to organize the content. Many people believe there is only one way to present an idea and follow their first impression. A more deliberate approach consists of listing several broad directions in which the content can be organized and then objectively analyzing each direction with the previously established criteria. This similarly can be done with layout decisions. Go through several iterations of the design process until grouping, sequencing, and labeling ideas are well flushed out.
5. **Prototype**: Make a rough draft with placeholder text. Sketch outlines on a white-board. Make font, color, and layout decisions. In other words, make a high-fidelity prototype. Once this is done, carefully evaluate the draft through the criteria. Ask the fundamental questions: Does this document meet the needs of the reader? Do ideas flow and are they developed in a logical way? Will the ideas contained in the document be understandable to the reader?
6. **Implementation**: Write the document.

Functional Specification

The functional specification is a special type of document designed to capture all the details of the project, serving as the central, most important, and often only communication vehicle between team members often separated by space and time. The functional specification is a document with extremely difficult usability challenges. These challenges come from several sources: the readers are a diverse and demanding group, the stakes are high for getting it right, and the thoughts the author is trying to communicate are complex and detailed.

Persona & Scenario

The specification is one of the rare times when the personas map directly to job titles.

	Persona	Scenario
Developer	Detail oriented, impatient	Needs an answer to a specific question
Tester	Very, very detail oriented	Enumerate test cases and make a test plan
Manager	Needs the big picture, does not like techno-babble	Collecting data from dozens of specs for a report
Peers in other teams	Just like you!	Looking for ideas, overlaps, and constraints
Localization	Foreign, intimidated by technology	Trying to find issues and estimate costs
Security	Developer, highly trained	Looking for issues through hundreds of specs

Observe how there are actually two classes of users: those who need to answer "What" and "Why" questions (Manager, Peers), and those looking for specific details (Developer, Tester, Localization, & Security). Unfortunately, these classes of users are completely different: the first group wants high-level overviews of the product and does not want to get bogged down in technology or details, the second group only wants the technology and details!

Criteria

Despite the two divergent sets of users, we need to have a single criteria set to satisfy all.

Priority 1

- The reader must be able to get an accurate high-level understanding of the basic composition of the feature in less than a minute (Efficiency & Mapping)
- The reader must be able to find the specific information he is looking for in less than a minute (Efficiency & Mapping)
- The reader must be confident that the information found is accurate and represents all the spec has to say on the subject (Trust).

Priority 2

- Information needed to answer a single question cannot be spread across multiple locations in the document (Visibility)
- The layout, headings, and structural elements must be similar to other specs in the organization (Familiarity)
- The document must not be painful to read or look at (Motivation)?

Design

Based on these criteria, there are several designs which we could pursue to address the technology/details duality discovered in the persona/scenario part of the design process.

Name	Separate documents
Summary	Create one spec for the managers, one for the technical people
Pros	Each document can be specialized to the specific persona
Cons	Failure to meet the trust criteria because the two documents could get out of sync.

Name	Label each section for who the content is for
Summary	Each heading of the spec will say "Managers" or "Testers" so each group of people can know what part of the spec is meant for them
Pros	Layout straightforward
Cons	A bit condescending thus not meeting the motivation criteria

There are benefits to both approaches. Wouldn't it be great if the spec could be organized so that the managers knew where to look in a document and it would be easy for them to scan through many documents at once, while at the same time not getting in the way of the technology-focused readers? In other words, take the best of the two designs?

Name	Two parts in a single document
Summary	The first page is designed to answer What and Why questions, the rest is designed to enumerate all the little details
Pros	All the information is in one document and each reader will know where to go
Cons	How exactly do we accomplish this? More detail is needed

Now, as part of the refinement process, we try to see how we can accomplish this broad design decision. We will start by enumerating all the information the managers need and then enumerate all that the technology people need:

Managers	**Technology-focused readers**
Executive summary	Lists of visual elements
Screen shots	Tables describing behaviors and actions
Persona & Scenario	Figures describing how elements relate to each other

It turns out that there is not an overlap between the two sets of readers. Thus the single document can have all the advantages of the two document design without the disadvantage of having separate documents. Also, the document can have the advantage of separately labeled sections without having to label the sections. Most functional specifications follow this design and have two parts: Page One and Detailed Design.

Page One

Page One is the part of the spec answering "Why" and "What" questions, giving the reader a broad overview of what the design is about, who it is for, and what problem it is trying to solve. While it is called "Page One," it seldom takes only one page. The parts of the page one spec are: executive summary, persona, scenario, and snap-shot. There may be other components as well.

Executive Summary This is lifted directly from the executive summary previously derived from early in the design phase. As mentioned previously, this is the first step of the design and the last. It should always represent the current thinking and is the last thing changed before the spec is completed.

It is very important that the value proposition is prominent in the executive summary. Some designers go so far as to have a separate heading drawing special attention to the value proposition. It should be very clear why this design, feature, or application is worth the time and effort to build!

Persona The persona section in a spec (sometimes labeled as "target audience") is a condensed version of the persona developed in the analysis phase. This section usually consists of only one or two sentences.

It is not uncommon for organizations to list segments of the target audience, characteristics of the target audience, or even just list the names for the personas in the specification. Whatever the format, the reader of the specification needs to know for whom the interface is intended.

Scenario As with the persona, the scenario component of the spec is a greatly reduced version of the one developed in the analysis phase. Only one or two sentences are needed here.

It is also common to list use-cases, user stories (for those following an Agile development methodology), case analysis, or some other variation.

Criteria A listing of the evaluation criteria, not the rubric. These should be prioritized of course. Some organizations prefer to list requirements in their specification, cross-referencing a features requirements with the overall project requirements. While these may be useful for determining whether the stake holder's needs are met, they do little to help us understand if our design is of an appropriate level of quality.

Snapshot We do not need or want a full prototype in the spec, only one or two representative screen-shots to help the reader visualize what the feature or product will look like in its final form. The spec author should constantly update and refresh the prototype picture as the design progresses and as the product is built so it always represents the most current thinking.

The snapshot helps executives, members of other teams, and other stake holders get the big picture. The detailed design part of the specification (described next) is filled with the minutia of detail that is often difficult for a reader to assemble into a holistic understanding of the system.

Detailed Design

The detailed design enumerates all the parts of the design in painful detail. There is absolutely no room for fluff, however. The second a developer detects fluff, he will tune you out and stop reading! Thus the document must be extremely concise. The usual way to accomplish the completeness and conciseness requirement is to make the detailed design section consist only of tables, lists, and figures. This is not a creative writing exercise!

Though the detailed design is only a single section in the spec, it accounts for the vast majority of all the pages. The typical real-world spec might be a hundred pages or so, the detailed design making up 90% or 95% of those pages.

The fundamental goal of the detailed design section is to be sufficiently detailed and unambiguous that any two developers will produce the same code if the design is carefully followed. This must be accomplished while also meeting the two priority one criteria for the detailed design section:

Variable	Criteria
Efficiency & Mapping	The reader must be able to find the specific information he is looking for in less than a minute. (Priority 1)
Trust	The reader must be confident that the information found is accurate and represents all the spec has to say on the subject. (Priority1)
Motivation	The reader should not be bored or find the reading process grading. (Priority 2)

The fundamental way to achieve this is two-fold. First, the structure or layout of the detailed design must be completely obvious. There are many ways to achieve this, each highly dependent on the exact nature of the feature or product the spec is describing. Second, the information must be presented in a way that is clear and concise.

The exact format of the detailed design, the layout of the section as well as the type of information presented, is completely up to the spec author. It only matters that the stake holders get the information they need and communication happens. That being said, a few tips are in order:

No fluff	The second a reader detects fluff, he will tune you out. Thus every single word must be relevant and contribute to the communication goal of the document.
Make frequent updates	The spec must be the last word on the current thinking of the design. As soon as any decisions are made to the design, update the spec immediately and communicate it with the stake holders.
Make it interesting	Give the reader a fighting chance to get through the document. Make it interesting. Make it visually pleasant. Include humor as appropriate, but not at the expense of clarity and professionalism.

Structure

Perhaps the best way to reveal the structure of the spec (and also the structure of the feature by extension) is by providing a map. Consider, for example, the mountain bike team web site described in previous chapters. There are several components to each page: the masthead, the content area, and the shopping cart. To communicate this to the reader, the following graphic is presented:

Figure 11.1 - A map of a home page where each region is denoted with a red rectangle

It is common to have the labeled regions in a map such as with hyperlinks, enabling the reader to jump to the section of their choice by clicking on the region of interest.

In a design that has a hierarchy of UI elements such as a web site, an application with multiple screens, or an application with many dialogs, it is often helpful to present a map in the form of a tree:

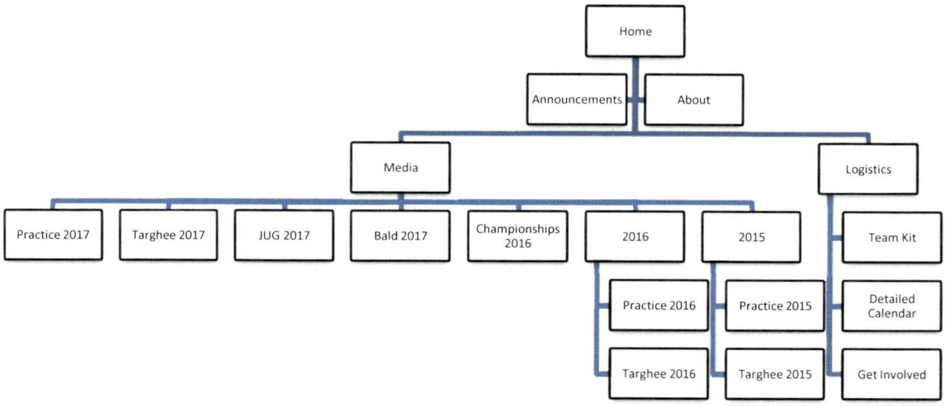

Figure 11.2 - A graph representing the organization of pages on a web site

Note that the same effect can be achieved with a standard table of contents.

Details

Most detailed design sections of specifications consist almost exclusively of tables, lists, and figures; there seldom is a single paragraph of descriptive text. Tables, lists, and figures facilitate skimming, are easy to recognize, and translate readily to code. As a rule, developers love figures, tables, and lists while seldom read large blocks of text.

Figures are helpful for displaying relationships that are difficult to describe with words. We could represent the same thing textually with a paragraph or two, but who wants to read that? Instead, a single marked-up figure is just as clear, more difficult to misinterpret, easy to find in a large document, and extremely fast to understand. One might specify the visuals for the OK button as the following:

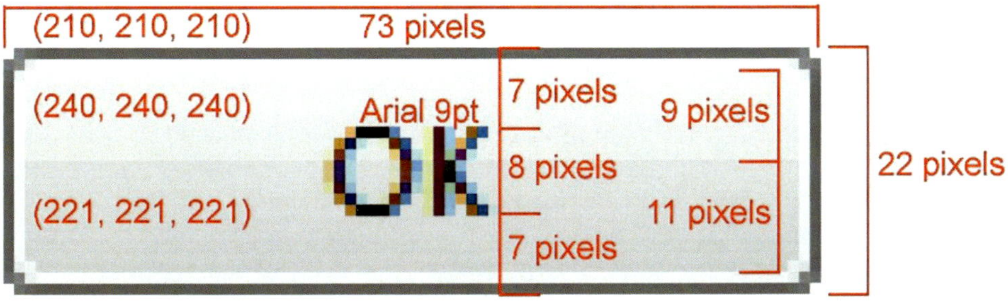

Figure 11.3 - A graphical representation of how a button is drawn

Tables are useful for conveying information that falls into a well-structured or repeating pattern. One example would be to illustrate the state of controls. In the following example, there is just two controls (OK and Cancel). Imagine how much information we could capture in a single page if there were a dozen controls?

Default	Disabled	Enter-focus	Tab-focus	Hover	Pressed
OK	OK	OK	OK	OK	OK
Cancel	Cancel	none	Cancel	Cancel	Cancel

Figure 11.4 - A collection of pictures representing the various states of the OK and Cancel buttons

We can also use tables to represent properties in an efficient and easy-to-understand way:

Property	Value
font-size	23pt
font-weight	bold
text-transform	uppercase
padding	0px
margin	0px
color	rgb(75, 75, 75)
border-bottom-style	solid
border-color	rgb(60, 60, 60)
border-width	3px

To express the layout of a dialog, web page, screen on a mobile application, or any other complex design, it is often necessary to present a heavily marked-up image. The following example is the main screen of a media player.

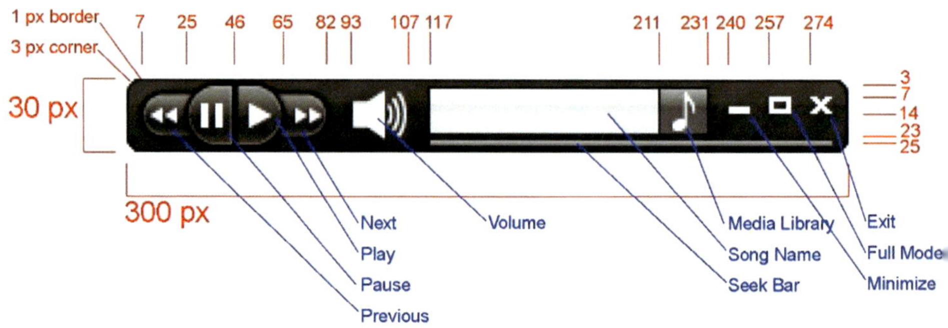

Figure 11.5 - Markup of a media player

Notice how the dimensions are clearly stated (300 pixels by 30 pixels), the horizontal position of the various controls is indicated, as are the vertical positions. There still remains much to be described, such as the colors, states of the various controls, the font used in the song name control, the behavior of the volume control, and the behavior of the seek bar.

Often we wish to express layout in relative terms because the various elements can move around on a page. This, too, can be described in a marked-up image:

Figure 11.6 - Markup of a UI element with complex formatting and alignment

Here we have several absolute sizes and positions. We also have relative ones (centered). Notice how the font attributes are drawn directly on the image. These could also have been done in a separate table below the figure.

Appendix

Frequently, out of necessity, an appendix needs to be added to the spec. In this appendix lies any information required by special clients or readers. Examples are:

Security The job of the security expert is to identify potential threats and make sure the user's assets are adequately protected. Adding an appendix for the security expert to answer these questions will make his job much easier.

Accessibility The job of the accessibility expert is to make sure people with physical or cognitive challenges are able to use the feature or product. Adding an appendix to enumerate all uses of sound, functionality to adjust text size, access to the operating system's accessibility functions, and potential dexterity issues will make his job much easier.

Internationalization The job of the internationalization and localization experts is to make the product work with multiple languages, multiple nationalities, and multiple cultures. Adding an appendix listing all strings and images to be localized greatly simplifies their jobs.

Product Support Product support and user assistant technicians provide help to the users after the product has shipped. Anticipating the questions they will answer and providing needed information greatly simplifies their job.

It is also common to provide other sections in the spec to make the spec-writer's job easier. Most of these sections are related to the following scenario: a team-member comes into the office of the spec writer and asks a question. Unfortunately, this question has been asked many times before. Rather than answer the question orally yet again, the spec writer decides to put the answer in the spec. Two ways to do this: a change history and an issues section in the spec.

Change History

One of the fundamental questions a reader of the spec asks when receiving a new version of the spec is: "what changed?" The spec needs to facilitate this question. Possibly the easiest way to do this is to add a Change History section to the appendix. This section includes a table where the first column is the date of the changes and the second describes the changes themselves.

An example is:

Date	Change
4/7/2018	First draft of page one
4/15/2018	Page one complete with placeholder graphics in the prototype
4/21/2018	High fidelity graphics from the graphic designer is incorporated
4/22/2018-a	Minor formatting changes for design review
4/22/2018-b	Adjusted persona based on feedback from design review
5/1/2018	First draft of detailed design
5/2/2018	Flushed out the home screen with rough graphics
5/9/2018	Home screen section complete with rough graphics
5/12/2018	Fixed minor typos and formatting errors
5/21/2018	Flushed out the chapter navigation section

One problem with this approach is that the old design ideas are not present. If a reader asks "what happened to the old design?" the spec will not answer that question with enough detail. To address this, we typically add a bone-yard section as well.

The bone-yard is where old ideas go to die. When a feature gets cut, a design gets revised, or a new graphic is used, the old design should not be deleted but rather moved to the bone-yard. For each entry, briefly describe why the change was made.

Issues

A final section that can be placed in the appendix is called "Issues." The purpose of Issues is to capture all the "stakes in the sand" and the big design decisions that were made in the design process. In other words, rather than answering the same questions again and again, the spec author explains the issue and includes a detailed description of the justification of the final decision. There are three advantages to this approach: people are directed to the spec as the main communication vehicle for all issues of the design, a written justification can be more detailed and carefully worded (read "more convincing") than an orally presented justification, and the spec author spends less time answering questions.

Example

The page-one component of the team mountain bike spec is the following. Notice a few things:

- The executive summary is at the top of the document immediately below the specification title.
- The personas and scenario are reduced from those developed earlier.
- The criteria are presented, but not the rubric.
- The snapshot shows a couple representative scenes from our prototype.

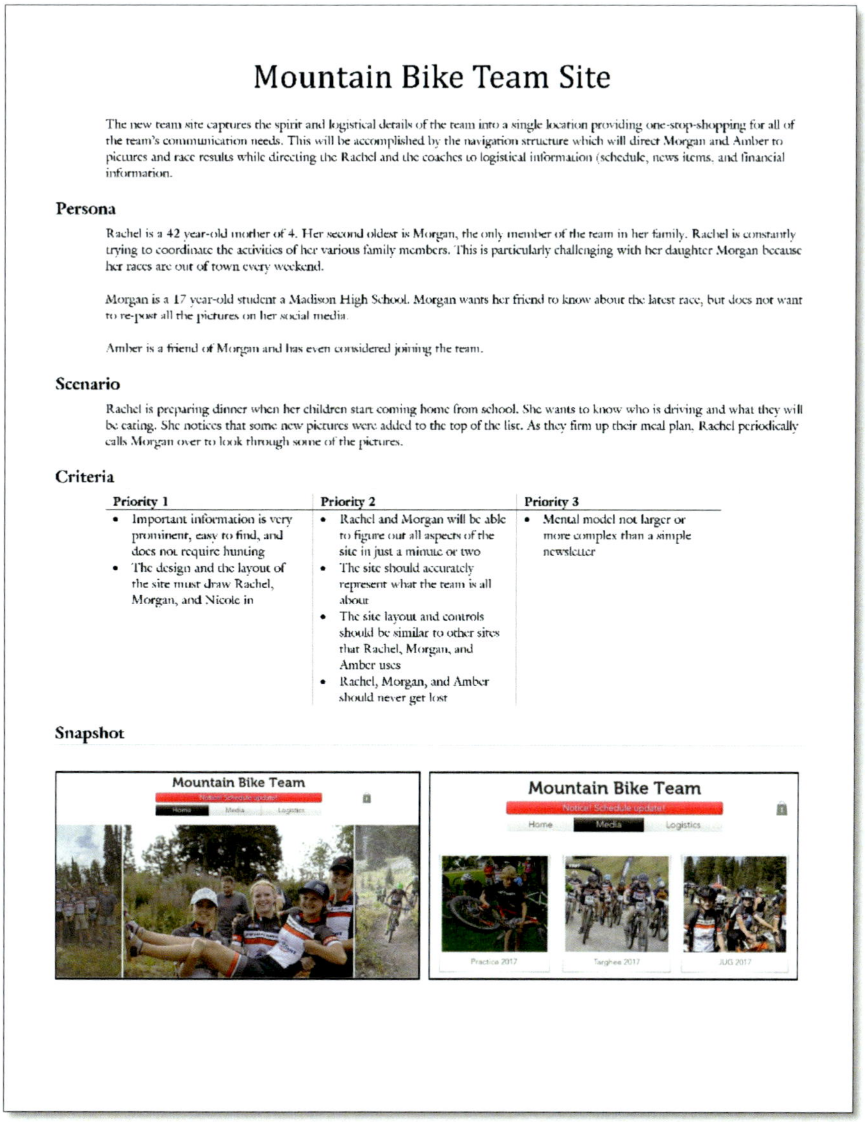

Mountain Bike Team Site

The new team site captures the spirit and logistical details of the team into a single location providing one-stop-shopping for all of the team's communication needs. This will be accomplished by the navigation structure which will direct Morgan and Amber to pictures and race results while directing the Rachel and the coaches to logistical information (schedule, news items, and financial information.

Persona

Rachel is a 42 year-old mother of 4. Her second oldest is Morgan, the only member of the team in her family. Rachel is constantly trying to coordinate the activities of her various family members. This is particularly challenging with her daughter Morgan because her races are out of town every weekend.

Morgan is a 17 year-old student a Madison High School. Morgan wants her friend to know about the latest race, but does not want to re-post all the pictures on her social media.

Amber is a friend of Morgan and has even considered joining the team.

Scenario

Rachel is preparing dinner when her children start coming home from school. She wants to know who is driving and what they will be eating. She notices that some new pictures were added to the top of the list. As they firm up their meal plan, Rachel periodically calls Morgan over to look through some of the pictures.

Criteria

Priority 1	Priority 2	Priority 3
• Important information is very prominent, easy to find, and does not require hunting • The design and the layout of the site must draw Rachel, Morgan, and Nicole in	• Rachel and Morgan will be able to figure out all aspects of the site in just a minute or two • The site should accurately represent what the team is all about • The site layout and controls should be similar to other sites that Rachel, Morgan, and Amber uses • Rachel, Morgan, and Amber should never get lost	• Mental model not larger or more complex than a simple newsletter

Snapshot

Figure 11.7 – Example of the first page of a specification.

The first page of the detailed design is the following. A few things to notice:

- The top of the detailed design is a graphic describing how the various components are organized in the specification.
- There are no paragraphs of text. Everything is a figure, a table, or a list.
- Every detail is described, giving the developer a complete picture of how the design is to look and behave.

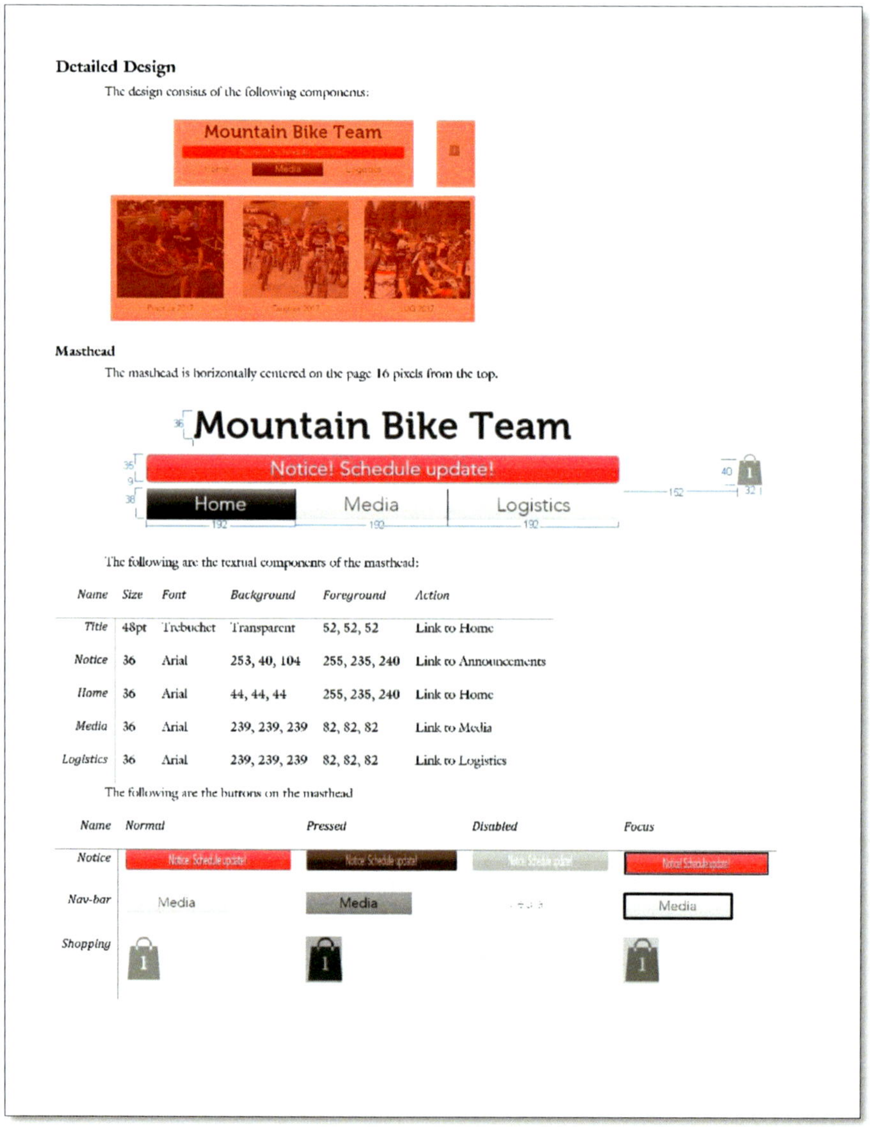

Detailed Design

The design consists of the following components:

Masthead

The masthead is horizontally centered on the page 16 pixels from the top.

The following are the textual components of the masthead:

Name	Size	Font	Background	Foreground	Action
Title	48pt	Trebuchet	Transparent	52, 52, 52	Link to Home
Notice	36	Arial	253, 40, 104	255, 235, 240	Link to Announcements
Home	36	Arial	44, 44, 44	255, 235, 240	Link to Home
Media	36	Arial	239, 239, 239	82, 82, 82	Link to Media
Logistics	36	Arial	239, 239, 239	82, 82, 82	Link to Logistics

The following are the buttons on the masthead

Name	Normal	Pressed	Disabled	Focus
Notice	Notice! Schedule update!	Notice! Schedule update!	Notice! Schedule update!	Notice! Schedule update!
Nav-bar	Media	Media		Media
Shopping				

Figure 11.8 – Example of the first part of the detailed design of a specification

Comprehension check

Quiz 1 **Parts of a Specification**

Describe in your own words what purpose each part of the specification is designed to accomplish

Quiz 2 **Résumé**

Create a draft of your résumé that adheres to the usability requirements of documents described in this chapter.

Problems

Problem 1 **Résumé**

Consider a résumé as a user interface design problem.

1. Who is the target audience for a résumé? List 3-4 and provide a couple sentences describing their essential characteristics.
2. For each persona from part 1, identify the scenario of the persona interacting with the résumé. Again, provide a couple sentences describing this scenario.
3. For each persona from part 1, identify a few of the evaluation criteria. Just a couple words for each will do here.
4. Identify the evaluation criteria for the résumé taking all the personas into account. Make sure you categorize the criteria (priority 1 through 3). Hint: most of the variables of usability will be relevant in this exercise.
5. Create a rubric describing how a résumé can be evaluated.

Problem 2 **Stopwatch app**

Write a spec for the user interface of the stopwatch app developed over the previous several lessons. Adapt the executive summary, persona, scenario, criteria, and prototype previously developed. Most of your effort should be spent on the detailed design section.

UNIT FOUR 4

Usability Studies

This consists of experimental evaluation tools and techniques allowing us to get a better understanding of how users react to various designs

12 Experimental Evaluation

As discussed earlier in the semester, there are three fundamental ways to evaluate a given usability design: heuristic evaluation, expert analysis, and experimental evaluation. Heuristic evaluation is the process of rating the design according to a set of standards or heuristics. This is typically done through the variables of usability. Expert analysis is performed by some acknowledged expert who gives his opinion based on experience and intuition. The final methodology, experimental evaluation, is the process of conducting a set of experiments designed to provide insight into targeted components of the design. The purpose of this article is to better understand why you may choose to perform an experimental evaluation, the type of information one can expect to gather, and the step-by-step methodology one follows when conducting such an evaluation.

What is Experimental Evaluation?

When you ask a layperson how to find the flaws in a given usability design, they will probably answer "just give it to the user and see how they like it!" The heart of experimental evaluation is really that simple: give the design to members of the target user group and look for the flaws. A well designed study will, of course, yield much better data than simply giving the product to the user and hoping for the best. However, as long as the person conducting the study is awake, useful data will come forth.

Think like a scientist

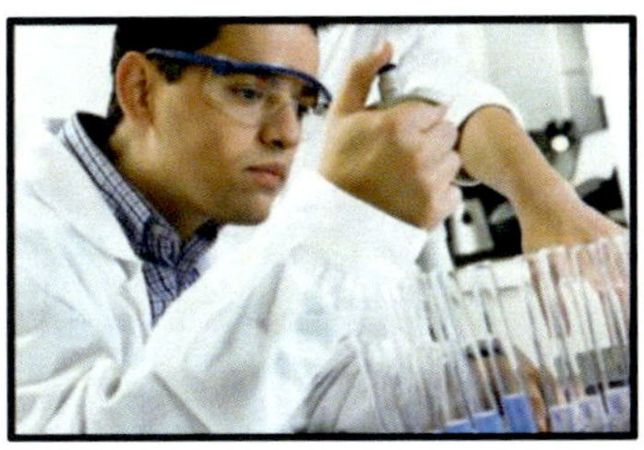

Figure 12.1 - Scientist

The key to creating a great experimental evaluation plan is to think like a scientist. Our prototypical scientist in this analogy is a chemist trying to uncover the properties of an unknown substance. Our scientist will follow the scientific method:

1. **Define the question**: What about the mysterious chemical are we trying to learn?
2. **Gather information**: Has anyone else attempted to solve the same problem? Look in the literature for similar studies.
3. **Form hypothesis**: Based on experience, intuition, and information gathered from the literature, he will form a hypothesis regarding the composition of the mysterious chemical.
4. **Perform an experiment**: Create an experiment to support or refute your hypothesis.
5. **Analyze the data**: Determine if the data collected supports or refutes your hypothesis.
6. **Draw conclusions**: Based on what you learned about this hypothesis, what does that tell you about your original question?
7. **Publish results**: Share what was learned with the larger community.

An effective usability engineer performs the same steps as the chemist in this example: he thinks of experiments he can conduct which will give insight into the problem he is attempting to solve. Each experiment should either support or refute a hypothesis or assumption about the design.

Strengths and weaknesses

Experimental evaluation can be the most accurate and insightful evaluation method if it is done correctly. The main reason for this is that we are directly measuring how the target user responds to the interface, as opposed to inferring this through indirect methods. Any large-scale design effort lacking an experimental evaluation component is thus at risk; there is no telling exactly how the user will react.

Figure 12.2 - Wristwatch

There are, however, several problems which need to be addressed:

- **Expense**: It takes a lot of time, effort, and money to conduct a thorough experimental evaluation. It is difficult to find good users, expensive to convince them to participate in the study, and painstaking to shift through the results.
- **"Why" questions**: Experimental evaluation is a useful tool for discovering what is good or bad about a design. However, it is often difficult to answer "why" questions: Why are the users so much less efficient than expected? Why could they not figure out this paradigm?
- **Bias**: Because we can never directly measure how the user reacts to the design, we are forced to infer based on observation. This inference is subject to a bias: the user may react differently because he knows he is being observed and the usability engineer may color his interpretation of the results based on pre-conceived opinion or belief.

Fortunately, with careful planning, these problems can be addressed in a carefully created usability plan.

Usability Studies

It turns out, a usability engineer follows exactly the same steps the scientist follows when creating an experiment: define the question, gather information, form a hypothesis, create experiments, create a usability script, conduct the experiment, analyze the data, draw conclusions, and publish the results.

Step 1. Define the question

While it seems like a trivial and obvious step, it is important early in the process to identify exactly what you are trying to find out. The usability study should have focus. There should be readily identifiable information that you are setting out to uncover. In other words, it is not OK to rely on serendipity.

Common questions that we try to answer include:

- How will members of the target audience react to the design?

- What are the most glaring defects?
- Will the design be good enough to satisfy the target audience?
- Is it better or worse than the competition?

Of course, we will follow a very different process when addressing each of these questions.

An example of this step would be a usability engineer tasked with making a summative evaluation (an evaluation near the end of the project to see if the product is on task) of a multi-function digital wrist watch. At this phase of the project, it is most important to identify defects and gauge overall impressions of the watch.

Step 2. Gather information

Before the study really gets underway, the usability engineer needs to be the expert in several areas:

Persona & Scenario Who is the target audience? The usability engineer needs to get into the head of the user, understand how they are likely to respond to a variety of designs, and understand their tastes and preferences.

Technology The usability engineer needs to develop a deep understanding of the technology being tested. He should have read the spec and worked with the design until it was thoroughly understood.

Design The usability engineer needs to be familiar with the paradigms the design is using and why the designers choose those paradigms.

Precedents Are there any applicable standards? Are there other designs attempting to solve the same problem? Will members of the target audience be familiar with these designs?

Back to our wrist watch example, the usability engineer carefully reviews the specification, paying special attention to the big design decisions and the target audience. He would then visit members of the target audience to develop a deeper understanding of their needs. He would thoroughly review the product itself as well as other watches of similar design the target audience may choose.

Step 3. Form hypotheses

A hypothesis is a belief or opinion the researcher holds regarding the design to be studied. There are several properties to a good hypothesis:

1. **Assertions**: Each hypothesis is stated as an assertion: something that is either true or false. This belief could go either way: "the user will be able to save a file without problem," or "the user will fail to find the 'Save' feature in an acceptable amount of time."
2. **Observable**: One can imagine a way to collect data to support or refute a hypothesis
3. **Relevant**: Resolution of each hypothesis will result in a better product. In other words, we do not bother stating things that don't matter.

4. **Directed**: Each hypothesis is directed towards helping the researcher address the questions identified in the first step.

How can we identify these hypotheses? There are two main tools we can use: the evaluation criteria from the design itself and the results of a heuristic (or even expert) analysis.

Each design should be based on evaluation criteria. If the researcher believes that the evaluation criteria do in fact represent the required success criteria describing a great design, then the researcher can use these criteria directly as his hypotheses. However, if the researcher believes that there are missing criteria or that some of the criteria are off-base, then those corrections can be the basis of the experimental hypotheses.

Consider again our wrist watch example. One of the evaluation criteria for this design is that the user will be able to start the stop-watch efficiently. The researcher can verify this with the following hypothesis:

> *The user will be able to time an event quickly and with little effort.*

Notice how this statement can be proven or disproven (more accurately: observations can support or refute this assertion). A good hypothesis has this property.

A second example would be the missing evaluation criteria. Our researcher realized that there is not a motivation evaluation criteria in the design. He also notices that little effort went into aesthetics. Theorizing this may be a shortcoming, he comes up with the following hypothesis:

> *The user will not like the watch and prefer less feature-rich designs based on aesthetics.*

Figure 12.3 - Wrist watch

Again, we can collect data to support or refute this hypothesis. If data is collected supporting this hypothesis, then the designer will be forced to take motivation into account with future iterations of the design.

A heuristic evaluation will give the researcher a set of defects which may exist in the design. The researcher can then use these conclusions as the basis for his experimental evaluation. If, for example, the researcher observes the set-time button is non-standard on the watch (Familiarity), he may turn that heuristic deduction into an experimental hypothesis:

> *The user will find the set-time button only after considerable searching and effort.*

Observe that, if this hypothesis turns out to be correct, the researcher will be able to make a strong "Why" assertion resulting in a convincing design recommendation.

Step 4. Create the experiments

Most of the hard work occurs in step four. Here the researcher creates a series of experiments designed to confirm or refute each hypothesis. These experiments are then combined into a single usability plan which will then be performed on members of the target audience.

Validity	The measure of how much an assessment instrument is measuring the right thing
Reliability	The measure of how accurate an assessment instrument is
Efficiency	The cost of obtaining a measurement

First the researcher comes up with several experiments each tied to a hypothesis identified in the previous step. When doing this, the researcher keeps in mind the three properties of any assessment instrument: **validity**, **reliability**, and **efficiency** (Worthen, White, Fan, & Sudweeks, 1999). The researcher addresses validity concerns by making sure the experiment measures what it is designed to measure: the hypothesis. Reliability concerns are addressed by minimizing errors or external factors in the design. In other words, how confident is the researcher that a second measurement of the same participant will give the same results? The final consideration is efficiency: how easy is it to obtain this measurement?

The researcher refines some of the hypotheses and removes others after several iterations. In the end, he should have a set of experiments that are likely to collect high quality data on every hypothesis. For example, consider the three hypotheses identified in Step 3:

	Hypothesis	Experiment	Interpretation
A	Easily start/stop timing	While meeting with the participant and with the watch on his wrist, ask him to time how long it takes for him to untie and retie his shoes.	If it takes less than 2 seconds, the design passes. If it takes more than 5, there are serious problems
B	Easily start/stop timing	Ask the user how quickly he can start then stop the timer. In other words, what is the minimum amount of time he can register on the timer?	Try to determine the limiting factor for the start/stop procedure.
C	Aesthetics	Ask the user what he likes about the design. Ask him what he likes least. Ask which watch he would prefer after showing him five comparable designs from other manufacturers.	The like/dislike questions have obvious interpretations. At least 50% of the users should choose the current design as their favorite
D	Set-time	Ask the user to set the time to the Mountain time zone.	Less than 10 seconds is good, 20 seconds or more is bad.
E	Set-time	Notify the user that the watch is too fast by 3 minutes and 10 seconds. Have him correct that time.	Look for sticking points, frustration, and the amount of effort required to perform the results.

Step 5. Create a script

The next step is to combine all the individual experiments into a single cohesive script. Note that, at a given point in the study, multiple individual experiments could be going on at the same time. It is required that all the experiments described above are incorporated. It is not required but highly desirable for the script to feel natural and non-contrived. Try to model the script off a mainstream scenario. Back to our wrist-watch example:

Step	Instructions	Experiment	Interpretation
1	Greet the user and explain what is to be studied		
2	Present the watch and put it in his hand. Ask him for his first impressions	C	Record results, look for likes and dislikes
3	Ask the user to put the watch on. Ask him how it feels. Does he like it or not?	C	Record results, look for likes and dislikes
4	Inform him the time-zone is set incorrectly. It is set to the Pacific time zone but it should be set on Mountain	D	Time him and record any comments or events
5	Inform him the watch has a timing feature. Ask him to time how long it takes him to re-tie his shoes	A	Less than 10 seconds is good, 20 seconds or more is bad.
6	Inform him the time is three minutes 10 seconds too fast. Ask him to set it to the correct time.	E	Look for sticking points, frustration, and the amount of effort required to perform the results.
7	Ask him how quickly he can set the start/stop timer button. What is the minimum time he can set on the stopwatch?	B	Try to determine the limiting factor for the start/stop procedure.
8	Ask him again if he likes the watch. What are the pros and cons	C	The like/dislike questions have obvious interpretations.
9	Show the user pictures of copies of comparable watches. Which would he prefer? Why	C	At least 50% of the users should choose the current design as their favorite
10	Thank the user and excuse him		

Step 6. Conduct the study

With a script, we are ready to conduct the experiment. This is done by taking the test plan to members of the target population and carefully recording the results.

Each participant should be a member of the target population. Notes should be taken reflecting data you collected that supports your claim they are members of this group. Without these notes, it significantly weakens the quality of the data you collect: how can you prove the data matters?

As you conduct the experiment, try to be cordial and friendly. The participant should enjoy the process and it should take the minimum amount of his time to help. While you value the data each participant gives you, he still retains the right to ask you to throw out your observations or to stop the study at any time.

The final guideline is to collect as much and as detailed data as you can. The more detailed these notes are, the more specific and targeted can your recommendations be.

Step 7. Analyze the data

Data analysis consists of comparing your experimental notes with the hypotheses. For a well-planned study, this is easy: the interpretation guidelines are built right into the study after all! However, when hypotheses are not present in the plan or when interpretation results are not well thought out, this can be a tedious and frustrating process. We will learn more about data analysis later this semester.

Back to our wrist watch example, we observed that the users were fast and efficient with both the stop-watch features and the set-time feature. This essentially supports the first hypothesis but refutes the third. However, none of the participants liked the design (third hypothesis). The average ranking was 4th out of 5 designs (only one design was less preferred) and they kept making remarks like "too big" and "too heavy." Also, two of the five participants noticed the edges were too sharp and were worried they might cut themselves.

Figure 12.4 - Wrist watch

Step 8. Draw conclusions

Based on your data and the hypotheses the data supports, the researcher next makes a series of recommendations for the designer. The goal of the researcher is to build a strong case for each recommendation. In other words, it should be nearly impossible for the designer to question your recommendations. Good recommendations have the following properties:

- **Based on analyses**: Every recommendation should be derived directly from the analyses derived from Step 5. Everything else is a matter of opinion.
- **Based on evidence**: Every recommendation should be based directly on experimental evidence gathered during the study. In other words, the researcher should always be ready to provide quotes, video clips, or other data to back up a recommendation.
- **Relevant**: All the recommendations should be directed towards making the product better. Inconsequential conclusions, no matter how accurate they might be, are not presented.
- **Do not design**: Finally, remember that the researcher is not the designer. He may point out defects but he is not allowed to make design recommendations.

Back to our wrist-watch example, we came up with three recommendations:

	Recommendation	Support
1	Make the watch smaller	All five participants observed the watch was bigger than the competition and they preferred smaller designs: • "Why is it so big" participant 2 • "I don't think I have ever seen a watch that big before!" participant 1 • "I think my wrist is too small for a watch that big" participant 3
2	Make it lighter	Four of five commented on weight. Quote: "It feels like I have a dumbbell in my hand" participant 3
3	Make it smoother	Two of five noted how sharp the corners are "Ouch!" participant 1 while in step 5 (tying his shoes)

Observe how each recommendation is based on a hypothesis and is further supported with direct evidence gathered during the study. There is no room for opinions in an experimental evaluation.

Step 9. Publish results

After the usability report is completed, it is presented to the design team. Almost certainly there will be defects in the design and opportunities for improvement. It is the obligation of the researcher to report them. However, it is equally important to be tactful.

Always remember that the design team typically has an emotional attachment to the project. This means they are typically sensitive to disrespectful, personal, or insulting remarks. At all points in time it is necessary to be tactful and positive.

> She didn't use the post formats selector once. She didn't even appear to see it.
>
> Apart from the video post, she was able to perform all of the tasks successfully with just the media modal and the editor.
>
> We really need to make sure that the "Insert from URL" feature inside the media modal embeds video URLS. We've seen users attempt to add videos this way in every single round of testing, and they are all very confused when a link is published, instead of the embedded video.
>
> The user selects multiple images and inserts them thinking she's added a gallery. I wonder how many users do this very same thing...

Notice how the results are posted in a blog and how the various members of the community are debating the implications of the test.

Example

Create a usability plan (steps 1-5) for the following dialog. Our study will focus on the “Effects” component of the dialog.

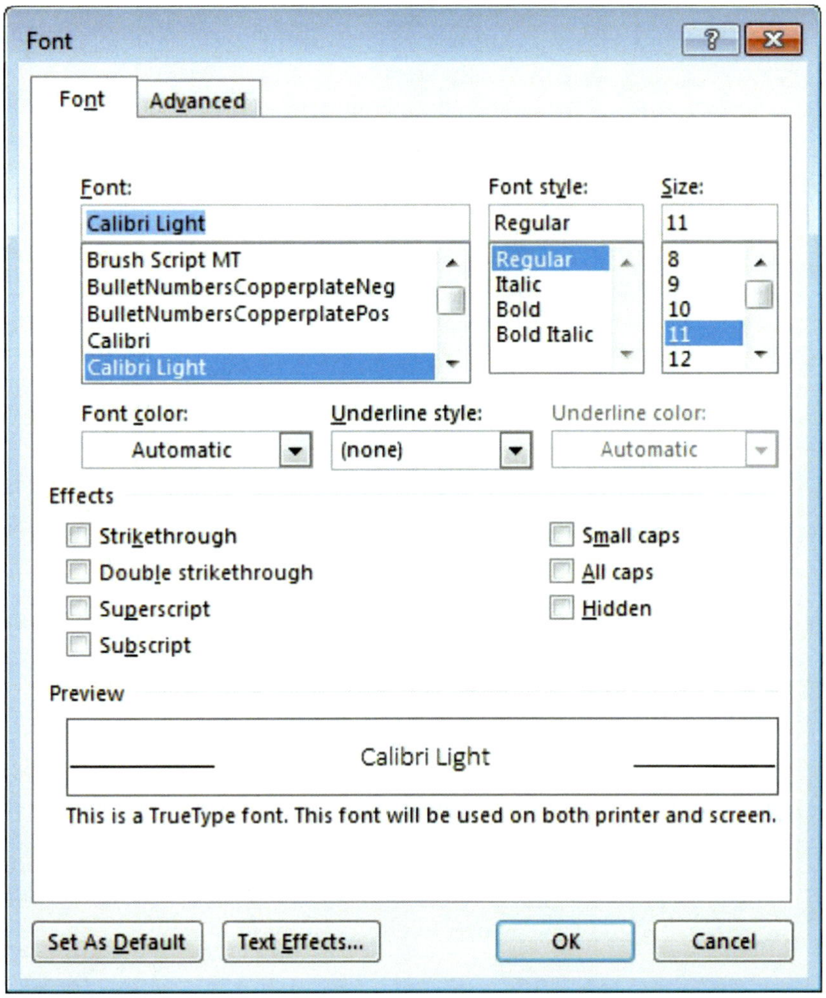

Figure 12.5 - Microsoft Word's Font dialog

Step 1 **Define the question**

We have two research questions that will need to be explored in this study. Each is equally important, as specified by our study sponsor.

- Are there any defects with the current design?
- Will the members of the target audience be pleased with the design?

Step 2 **Gather information**

Based on data that was collected in the design of this dialog, we discover the following information:

- Target audience is "all computer users." This was sub-divided into two groups: information workers who make heavy use of the dialog, and novice users who are on the beginning slopes of the learning curve.
- There are three use-cases: selecting a formatting, verify that a given formatting property is applied, and see what formatting options are available.
- The technology is a modal dialog in a desktop application using the default controls that are available to all dialogs on this system. There are no custom controls in this portion of the dialog.
- In speaking with the designers, we learn that an important goal was to reduce the perceived complexity of the interface. They felt there was a significant risk that the user would feel overwhelmed. All options were therefore presented in the simplest format possible. The evaluation criteria were:

  Priority 1:

  1. Visibility: The user can see all the character properties on a given run of text
  2. Visibility: The user can see all the character formatting options that are available
  3. Efficiency: The user can quickly make a formatting selection

  Priority 2:

  4. Mapping: The user can understand the relationship between the properties so he/she can make an informed decision
  5. Trust: The interface appears to the user as non-intimidating and safe

  Priority 3:

  6. Motivation: The interface looks to the user professional
  7. Motivation: Making a formatting selection is fun/rewarding

- Since all the controls in this portion of the dialog are check-box controls representing Boolean data, we understand that [x] implies the setting is applied to the underlying text and [ ] implies that the setting is not applied

Step 3 **Form hypotheses**

Looking at our three use-cases, we can identify two important hypotheses:

> *The user will be able to understand which formatting effects are applied to his/her text selection.*
>
> *The user will know how to apply superscript.*

Next we move to the evaluation criteria. We have a few hypotheses relating to that:

> *The user will not know what the term "Small caps" means.*
>
> *The user will think that "Hidden" is related to "Small caps" and "All caps."*

Finally, after working with the dialog for some time, we notice a few other things. First, we notice that the "Superscript" and "Subscript" controls work more like a radio group: they are mutually exclusive. The same is true with "Strikethrough" and "Double strikethrough" as well as "All caps" and "Small caps." Perhaps this might be confusing to the user. After spending some time with this, our usability engineer decides that they are not a problem but decides to test it just in case:

> *The user will not be confused by the toggle nature of the Superscript/Subscript controls.*

Based on these five hypotheses, the following set are created:

1. The user will be able to understand which formatting effects are applied to his/her text selection.
2. The user will know how to apply superscript.
3. The user will not know what the term "Small caps" means.
4. The relationship between the groups of related controls will not be confusing to the user.

Step 4 **Create the experiments**

The next step is to create at least one experiment for each hypothesis.

	Hypothesis	Experiment	Interpretation
A	Understand	Create a document with a wide variety of text effects. Each effect and all combinations of effects are present in the document. Have the user mark on a sheet of paper which effect is present at each location in the document.	If the user makes no systematic error (such as always misidentifying small caps as all caps), then the hypothesis is supported.
B	Understand	Create a collection of text effects in many combinations. Take screen-shots of the effects section of the dialog in various states. Ask the user to match the screen-shot with the text effect.	If there are no systematic errors, then the hypothesis is supported.
C	Apply	Create a document with the words "superscript", "superscript-strikethrough," and others. Ask the user to apply the formatting to match the labels.	If there are no systematic errors, then the hypothesis is supported.
D	Apply	From a screen-shot of the dialog, ask the user to describe what each text effect would look like when applied to a document.	Any deviation from correct answers would refute the hypothesis.
E	Labels	List the text effect labels on a sheet of paper. Ask the user to describe what each does.	Any deviation from correct answers would support the hypothesis.
F	Labels	Ask the user to explain the difference between all caps and small caps.	Any deviation from correct answers would support the hypothesis.
G	Groups	Ask the user to make text that is both superscript and strikethrough. Ask the user to make text that is both superscript and subscript.	If the user claims that it is impossible before attempting to apply superscript and subscript, or if the user comments after attempting it that it should not be possible, then the hypothesis is supported.
H	Groups	Place the labels of the effects on individual scraps of paper. Ask the user to organize them. Next ask the user if the organization as presented in the dialog makes sense to him/her.	There should be three groups of 2 labels each with "Hidden" being alone. Anything else would refute the hypothesis.

Step 5 **Create a script**

Step	Instructions	Experiment	Interpretation
1	Greet the user and explain what is to be studied		
2	List the text effect labels on scraps of paper and ask the user to describe what each does	E	Any deviation from the correct answers would support the hypothesis
3	Ask the user to explain the difference between all caps and small caps	F	Any deviation from correct answers would support the hypothesis
4	Ask the user to organize the labels into groups	H	There should be three groups of 2 labels each with "Hidden' being alone. Anything else would refute the hypothesis
5	Bring up the dialog and ask the user if the organization as presented in the dialog make sense to him/her	H	Any negative comments would refute the hypothesis
6	Create a document with the words "superscript", "superscript-strikethrough," and others. Ask the user to apply the formatting to match the labels	C	If there are no systematic errors, then the hypothesis is supported.
7	Ask the user to make text that is both superscript and strikethrough. Ask the user to make text that is both superscript and subscript	G	If the user claims that it is impossible before attempting to apply superscript and subscript, or if the user comments after attempting it that it should not be possible, then the hypothesis is supported.
8	Create a document with a wide variety of text effects. Each effect and all combinations of effects are present in the document. Have the user mark on a sheet of paper which effect is present at each location in the document	A	If the user makes no systematic error (such as always misidentifying small caps as all caps), then the hypothesis is supported.
9	From a screen-shot of the dialog, ask the user to describe what each text effect would look like when applied to a document.	D	Any deviation from correct answers would refute the hypothesis.
10	Create a collection of text effects in many combinations. Take screen-shots of the effects section of the dialog in various states. Ask the user to match the screen-shot with the text effect	B	If there are no systematic errors, then the hypothesis is supported
11	Thank the user and excuse him		

Comprehension check

Quiz 1 **Scientific Method**
From memory, list the steps of the scientific method

Quiz 2 **Experimental Evaluation**
From memory, list the steps of the experimental evaluation process.

Quiz 3 **Weaknesses of Experimental Evaluation**
List all the weaknesses of experimental evaluation

Quiz 4 **Deriving Hypotheses**
From where does the usability engineer derive the set of hypotheses used in the experimental evaluation process?

Quiz 5 **Parts of a Hypothesis**
List all the components of a well-formed hypothesis

Quiz 6 **Recommendations**
List all the properties of a good usability study recommendation

Problems

Choose a mobile application that you use every day. We are going to create a usability study for this application. This study will answer two questions:

- Does the current design serve the target audience?
- What is wrong with the current design?

Problem 1 **Step 2: Gather Information**
For the this mobile application, identify:

1. Persona & Scenario
2. Technology
3. Design
4. Precedents

Problem 2 **Step 3: Form Hypotheses**
Identify a collection of hypotheses about the application. Each hypothesis should be related directly to a variable of usability.

Problem 3 **Step 4: Create the experiments**
Identify a collection of experiments to test each hypothesis.

Problem 4 **Step 5: Create the script**
Create a script to capture the experiments.

13 Data Collection

A critical part of any user interface design or evaluation effort is data collection. We must have a high-fidelity understanding of the target audience, their task, and how they react to various aspects of the design. This chapter will help us better understand how to perform this critical task more effectively.

The usability engineer has two goals when collecting data: to ensure the participants' needs are met and to gather the highest quality data possible. It is easy for the engineer to be overly focused on the data she is collecting and forget the needs of the participants. However, if the participants find the data or the data collection techniques to be unsatisfactory, then the study might be worse than a loss. It might be a disaster.

There are three things to carefully consider when conducting a usability study: the participants, the data, and the specific interview techniques that will be employed.

The Participants

Figure 13.1 - Participant and usability engineer

We must always remember to keep the needs of the participants foremost in our minds when conducting a usability study. The usability engineer should always be ready to end an interview or throw away data if the participant finds any part of the study to be objectionable. This can be difficult to do; after all the data is valuable to the researcher. In order to minimize the chance of this happening, the researcher needs to take the following precautions:

1. Carefully plan the study so the participants are inconvenienced in the minimal possible way. Any interview, no matter how short, is still an inconvenience to the participant. The researcher should try to make the time seem like a "good deal" for the participant. This may include paying them for participation (with pizza or candy, typically) and to make the interview itself as fun as possible.

2. Be up-front with the participant about how the interview is to be conducted. A big part of this is to describe the purpose of the study and give a full disclosure of how the data will be used. There should be no surprises.

3. Make any data collection obvious. For example, if audio is recorded, place the recording equipment in a prominent location during the interview (such as on the table). It is often helpful to allow the participant to review the data at the end of the interview.

4. Have the participant sign a permission form before any data is collected. The form can be quite simple. Please see the model on the following page for an example.

To: Prospective usability study participant

Subject: Usability study for a mobile application media player

You are invited to participate in a usability study to help us find the defects in a piece of software. This study will consist of the usability engineer walking through a series of steps designed to represent a typical use-case of the software being studied. You will be asked to interact with the software and give your feedback as to what is good or bad about the current design.

During the course of this interview, two kinds of data will be collected. First, the usability engineer will take notes of the key events that transpired during the course of the study. You will be given the opportunity to review these notes at the end of the study. Second, the usability engineer's computer will record your interactions with the software. This will include mouse movements as well as how the software responds to your input. Again, you will be given the opportunity to review the resulting movie at the end of the interview if you desire.

Participation in this study is voluntary. If at any point in time you would like the interview to end or the data to be erased, please indicate it to the usability engineer. Thank you in advance for your participation.

Signed: Sam S. Student

The Data

Of course, the whole point of conducting a usability study is to collect data. In order to be effective in this regard, it is essential for the usability engineer to be intimately familiar with all aspects of the study. She needs to know how to minimize experimental bias. Finally, she needs to choose the best data recording technique so the details of the interview can be reconstructed as accurately as possible.

Be familiar with the study

If an interview is conducted before the usability engineer is fully prepared, it is likely to take more of the participant's time, include mistakes that could invalidate some of the data collected, and overlook key observations a better prepared interviewer would make. In order to prevent these events from occurring, it is essential for the usability engineer to thoroughly familiarize herself with all aspects of the study.

First, the usability engineer needs to be familiar with the target audience. She should know what to expect from each participant, an important part of keeping each participant in their comfort zone. She should be familiar with their special needs or issues. Finally, a well prepared engineer will be able to predict how a participant will respond to a given situation even before the situation presents itself.

The usability engineer needs to be familiar with the technology. This includes the subject of the study as well as any technology used to gather information. Usually this can be accomplished with a series of trial run-throughs. In the end, if the plan runs amuck, can you recover and still collect the data you need?

Finally, the usability engineer needs to know the usability plan thoroughly. Is the plan itself solid or will some re-working be necessary? Can the plan be adapted if the situation demands it? Can the study be conducted without prompting or reading from the cue-sheet? When the participants, the technology, and the plan are completely understood, then the study can begin.

Minimize bias

It is impossible to make a measurement without affecting the thing being measured. This is true when measuring atoms or measuring people. Our job when conducting a usability study is to minimize the amount of experimental tainting and to ensure the data is the truest representation of what the participants really think.

The best way to minimize bias is to minimize the interactions of the usability engineer with the participants. Often this can best be done by getting the participant to forget he is being observed. For example, a user being observed in his own office, doing his own work, is probably a truer representation of how he uses the software than when the same user is brought into a lab and asked to perform a series of tasks. There is a flipside to bias, of course. The less interaction the engineer has with the participant, the less likely the type of data the engineer is seeking will be collected.

Collect the data

Figure 13.2 - Data collection

The final thing to be considered when preparing to conduct a usability study is how the data will be collected. In the end, the goal is to collect enough data that the important details of the interview can be reconstructed for later analysis. In a lab setting, it is often possible to collect multiple types of data for a single event. For example, the computer could record all the interactions between the user and the software, a video camera can record the participant's reactions to the study, and the usability engineer can take personal notes reflecting impressions or events likely to be otherwise missed.

When making digital recordings of a participant, make sure the participant is fully aware of the recording, and knows how the data will be used. It is often helpful to allow the participant to review the data on request. When this permission is granted, screen capture software or even your laptop's microphone can easily satisfy this part of data collection.

When making hand-written notes of an interview, a tradeoff needs to be made: the more detailed the notes, the more the note-taking process will be a distraction to the participant and the more experimental bias is likely to result. When notes are not taken during the study or when only key events are recorded, the risk exists than important data will be missed. Where do you draw the line? This is a judgment call that must be made by the usability engineer.

The Techniques

There are many techniques that can be used when conducting a usability study. You should not look at these techniques as "my favorite" or "this is the best," but rather as a collection of tools in a toolbox which may be useful to solve a given problem. In other words, it is a good idea to familiarize yourself with all these techniques so, when the situation arises, you can apply them effectively. The most common techniques are interview, questionnaire, thinking aloud, ethnography, cooperative, and card sort.

Interview

If you were to ask most people how to find out what is wrong with a given design, they would probably respond "just ask the users." This is the essence of the interview technique: asking the user questions and carefully recording their response. The quality of data gleaned from an interview is directly related to the quality of questions posed to the user and the willingness (and ability) of the user to respond.

When to use The interview technique is best used when a great deal of flexibility is needed in the types of data collected. Interview techniques allow the conversation to take unexpected turns in response to revelation and insight given by the participants.

When to avoid Conducting interviews is a time-consuming activity, requiring a great deal of effort to conduct and even more to analyze the resulting data.

What is needed Interviews require very little in terms of physical resources. One only needs a cue sheet for the interviewer and the ability to record the conversation.

Tips and guidelines While any interview is likely to provide insight into the user and the interface being studied, a few things can greatly increase the quality of the data:

- Appear friendly and open-minded to the participants. The more the participant likes you, the more she is likely to give you good data.
- Make data recording subtle. People tend to clam-up when they see a recording device or watch you make notes on a sheet of paper.
- Ask open-ended questions. Your goal is to talk less than the participant.
- Ask only questions they can answer. Put questions in their words and speak from their context. If the participant does not know how to answer a question, it is a wasted question!

Interviews represent a component of virtually every usability study.

Questionnaire

The questionnaire technique is similar to the interview technique except the interviewer is not present; all the interactions occur asynchronously through a paper or electronic form. This is accomplished by simply writing down a number of questions on a sheet of paper or in a web survey and distributing it to the target population. Unfortunately, unlike an interview, it is impossible to adjust the questions in the middle of the study. If a question is confusing or does not lead to the desired responses, there will be a gaping hole in your data. Therefore it is very important to get the questions right before distributing the questionnaire.

When to use Questionnaires are a great tool to get a large amount of data with little effort/cost. If you wish to get enough data to make statistical conclusions, then questionnaires are your only viable alternative.

When to avoid Typically two types of people respond to questionnaires: those who are very happy and those who are very upset. As a result, data tends to be biased and polarized.

What is needed An important questionnaire consideration is the delivery mechanism. How are you going to get this questionnaire in front of the right people? In the past, postal-mail surveys were the delivery mechanism of choice. Today, internet-based surveys are quite common.

Tips and guidelines It is relatively difficult to come up with a questionnaire good enough to provide useful data. To help with this process, the following guidelines are provided:

- Carefully vet the questions with co-workers and friends. Agonize over every single word in an effort to minimize confusion.
- Ask each question two times in slightly different ways. This will allow you to cross-reference answers.
- Ask open-ended questions. Make sure to give the participant a chance to express herself.

While a questionnaire may be an inexpensive and easy tool to gather a large amount of data, you should almost never use it in exclusion of other techniques.

Thinking aloud

Thinking aloud is a collection of techniques designed to discover what mental and emotional processes are going through the participants head when interacting with the system. This is accomplished by simply asking the user what he is thinking at various moments in the usability study. There is one problem with this technique: people have a tendency to go quiet the very moment when they are thinking the hardest!

When to use Every time the user is interacting with a prototype or an actual piece of production software, consider using the thinking aloud method.

When to avoid Just talking about a problem has a tendency to change the thought process. People often perform much better while thinking aloud than they do when working silently.

What is needed Nothing. No setup or planning is required.

Tips and guidelines To maximize the effectiveness of this technique, consider the following:

- Do not be overly repetitive. If you ask "what are you thinking" every 30 seconds, their response will likely become "I am thinking of how annoying you are!"
- Prompt the participant only occasionally, usually when they have gone silent for a few seconds or are obviously thinking.

Ethnography

Ethnography is the process of observing users in their "natural setting," using the product as they normally do in their own home or office. This technique gives us rich data about how the user normally interacts with the software and provides rich persona and scenario data.

When to use This is highly useful at the end of a product cycle when it is possible to get production or pre-production products in front of the user. Great for understanding the user and the user's context.

When to avoid The danger exists that the user can be tempted to expose confidential data or, if pre-production software is used, the possibility exists that harm will come to the user or his data. Make sure to keep the participant's needs foremost in your mind. It is also hard to get the information you want or need!

What is needed The only thing that is needed is permission to sit in someone's office while they are working and a notepad to take notes.

Tips and guidelines Insightful data will almost certainly result from ethnography studies. The most important tips center around catering to the needs of the participant:

- Try to be invisible. Only interrupt the user when you absolutely need to clarify something.
- Make the inconvenience worth-while for the participant. Bring them a gift or some other compensation in exchange for their time.

Remember you are a guest in the participant's space. Try to make your visit as non-invasive as possible.

Cooperative

Cooperative evaluation is the process of the interviewer and participant teaming up to uncover what is good or bad about a given design. In other words, most interviewing techniques pit the interviewer against the participant (sitting on opposite sides of the table so to speak). The cooperative method involves the interviewer sitting beside the participant, working together to uncover issues. It usually begins with an introduction like "we know there are problems with this design but are having trouble uncovering them. Maybe if we work together we can find them all."

When to use The cooperative technique is most useful when the participant is having trouble opening up. It can also be useful when trying to overcome social hurdles. For example, some participants find it difficult to criticize, thinking they will hurt the feelings of the designer.

When to avoid The more the interviewer injects his opinions, the more the data will become a reflection of the interviewer's thoughts rather than the participants. There is a huge potential for experimental bias.

What is needed No extra planning or material is needed to conduct this style of interview. Often the interviewer can slip into the cooperative mode when, mid interview, it becomes evident that the participant is having trouble opening up.

Tips and guidelines While it is fairly easy to conduct a cooperative-style interview, there are a few guidelines:

- Try to keep your opinions to yourself. Because experimental bias is such a problem, the interviewer needs to speak as little as possible so as to avoid influencing the participants thinking.

- Remove blame. Most people think that flaws in a design are the result of a mistake, the uncovering of which could get the designer in trouble. It is often helpful to emphasize that the designer is not being criticized, it is the design. The only way for the design to improve is to uncover the issues. Thus the study is not about getting people in trouble but rather helping the designer reach his potential.

Card sort

A card sort is a technique designed to uncover how a user or group of users think data or functionality should be organized. The technique is simple in concept but rather sophisticated in the implementation. The basic idea is that the usability engineer creates a list of cards (though software methods are also commonly used) on which are listed the various user interface elements for a design problem. The participants then organize the cards into categories according to how they think things should be grouped. There are many types of information that can be gleaned from such a study: the user's mental model of how the system should work, labels or feature names which are not self-explanatory, and the participant's thought process as they perform the sort. The last item is the gem of card sorting: people have a tendency to talk aloud when sorting. This is particularly true when working in small groups. (Hudson, 2013)

When to use Early in the design phase when trying to uncover the existing mental model of the target audience. Late in the design phase when verifying that the labels of various interface elements is clear and unambiguous.

When to avoid The card sort is a specialized technique that cannot be applied to all problems.

What is needed A paper card sort requires a collection of index cards with one interface item represented on each. Computer-mediated card sorts are also possible, allowing participants to conduct a card sort remotely.

Tips and guidelines The most important tip for doing a card sort is to stay out of the way: let the participants work through the problem un-interrupted.

- Discretely record oral comments. As the participants work through the problem and think aloud, pay special attention to what they say. These unguarded moments represent some of the best data we can gather regarding what people think. As usual, make sure that the participants agree to this data collection before the sort begins.

- Use the right number of items. A card sort with a dozen or fewer items is trivial and unlikely to yield interesting data. A sort of a hundred items is not fun, tempting the participants to "get through it" rather than sharing with you what they really think.

Parting Thoughts

While considerable experience is often required to get the most out of every interview situation, anyone can collect data. As a seasoned interviewer once said "the only way to not collect useful data in an interview is to not have the interview in the first place."

Comprehension check

Quiz 1 **Precautions**
What precautions should be taken to ensure the participant's needs are met during a usability study?

Quiz 2 **Amount of Data**
How much data should be collected during a study?

Quiz 3 **Destroying Data**
Under what circumstances should the usability engineer destroy valuable data collected during a usability study?

Quiz 4 **Technique Names**
From memory, list and define the six data collection techniques.

Problems

Problem 1 **Interview**
Consider the following situation:

> Your team is about to begin the creation of a new web site for the local high-school track team. You are tasked with getting to know the parent stakeholders.

Create a participant permission form, create an interview question cue-sheet, conduct an interview of a parent to a high-school athlete. You can simulate this experience by asking a member of the class impersonate a parent to a student athlete.

Problem 2 **Questionnaire**
Consider the following situation:

> Your team has just finished creating a web site for the local high-school track team. You are tasked with discovering how the student athlete stakeholders will respond to the new site.

Create an online questionnaire.

Problem 3 **Thinking Aloud**
Consider the following situation:

> You have just completed the first draft of a web site for the local high-school track team. The head coach is an important stakeholder and you have a meeting with her.

Create a questionnaire on a sheet of paper. Give that questionnaire a student athlete for a high school track team. You can simulate this experience by asking a member of the class impersonate a student athlete. Simulate an online questionnaire with pen and paper.

Problem 4 **Ethnography**

Consider the following situation:

> The final version of a new web site for the local high-school track team has been completed. The coach just posted the latest results from the week-end track meet as well as about 40 pictures.

Conduct thinking aloud study on a coach for a high school track team. You can simulate this experience by asking a member of the class impersonate a high school track coach. You can simulate a completed web site by searching the web for a site that already exists.

Problem 5 **Cooperative**

Consider the following situation:

> A partial working system of a high school track team web site has been completed. While a great deal of work still needs to be done, the site is basically up and running. You would like to know if there are any glaring defects before continuing with the development efforts.

Conduct a cooperative study of the web site inviting on the student athlete's parents to walk through the design with you. You can simulate this experience by asking a member of the class impersonate a student athlete's parent. You can simulate a completed web site by searching the web for a site that already exists.

Problem 6 **Card Sort**

Consider the following situation:

> You are the member of a team tasked with building a web site for the local high school track team. The customer (the head coach and half-dozen assistant coaches) have just given you two dozen items that need to be in the site. These include: the yearly schedule, uniform care instructions, waivers and permission slips, descriptions of the various events, pictures of last year's meets, fees, sponsorships, record times/distances for each event, contact information for the coaches, practice times and locations, current weather and forecast, pictures from the recent track meet, details about the various track meets, links to training tips for the various events, stretching instructions, bad weather cancelation policy, and practice/training schedules.

Conduct card sort of any five stakeholders for a high school track team: student athletes, parents of student athletes, coaches, and prospective athletes. You can simulate this experience by asking a member of the class impersonate a stakeholder.

Problem 7 **Web Site**

Which data-collection technique should be used in the following situation:

> Your team has produced a new web site for the local high school. The site consists of a couple dozen pages addressing a wide variety of topics: academics, activities, athletics, and administrative. You have been tasked with studying the new design and am concerned that the organization structure will not be intuitive or efficient for the target audience.

List each of the six interview techniques. Briefly (in a sentence or two mention why a given technique would or would not be useful in this scenario. If it would be useful, describe how to carry it out.

Problem 8 **Desktop Application**

Which data-collection technique should be used in the following scenario:

> Consider the drawing application that comes bundled with the operating system you use for your laptop or desktop computer. You are part of a team that has just re-designed this drawing program. We have a new design early in the development process and want to get a feel for how the users will react to it. We would also like to uncover defects in the design while there is still time to fix them.

List each of the six interview techniques. Briefly (in a sentence or two mention why a given technique would or would not be useful in this scenario. If it would be useful, describe how to carry it out.

Problem 9 **Mobile Application**

Which technique should be used in the following scenario:

> As part of their regular investment in the mobile space, a large social media provider is contemplating a major redesign of their flagship app. Before the design process begins, they want to get a better understanding how their customers use the existing design.

List each of the six interview techniques. Briefly (in a sentence or two mention why a given technique would or would not be useful in this scenario. If it would be useful, describe how to carry it out.

14 Qualitative Analysis

There are fundamentally two types of experimental data: quantitative and qualitative. Quantitative data, defined as data in the form of numbers that can be counted (Oxford University Press, 2010), is possibly the most familiar data type to the average person. Common examples of quantitative data include polling numbers during an election, your grade on an exam, and the amount of time it takes a user to complete a given task. Quantitative data is typically manipulated in a spreadsheet and analyzed with statistics. Common statistical procedures used on quantitative data include the mean (the average value in a set of data), standard deviation (the amount of variance in the data), and the t-test or ANOVA (the degree in which conclusions drawn from a sample of the population can be extrapolated to the entire population). Engineers typically like quantitative data because they are comfortable working with numbers.

Qualitative data, on the other hand, is defined as data in the form of words or textual descriptions. Qualitative data typically resists application of statistical procedures because they are not in numeric form. Common examples of qualitative data include answers to questions, descriptions, and impressions. Qualitative data is typically captured with an audio/video recording device or in a text document. Social scientists typically like qualitative data because it captures the richness of individual experiences and observations. They are the source of "well grounded, rich descriptions and explanations of processes in identifiable local contexts" (Miles & Huberman, 1994, p. 1). To analyze qualitative data, we use tools such as pre-structured case outlines, coversheets, coding, and vignettes.

Overview

Qualitative data analysis in the context of a usability study begins before the first participant is brought into the lab. The researcher first thoroughly familiarizes himself with the product being analyzed, the target audience represented by the persona, and the usability plan. It is typically a good idea to pilot the study with a couple convenient participants (coworkers, roommates, and friends) to make sure there will be no surprises.

Perhaps this process is best explained by example. Consider a usability plan analyzing the print properties dialog box for a new printer.

Pre-structured case outline Identify what answers we are looking for in the study

Coversheets Each interview is summarized with a coversheet

Coding Tag the interview so all the data about a given topic can be collected

Vignette Summarize groupings of similar data

Pre-structured Case Outline

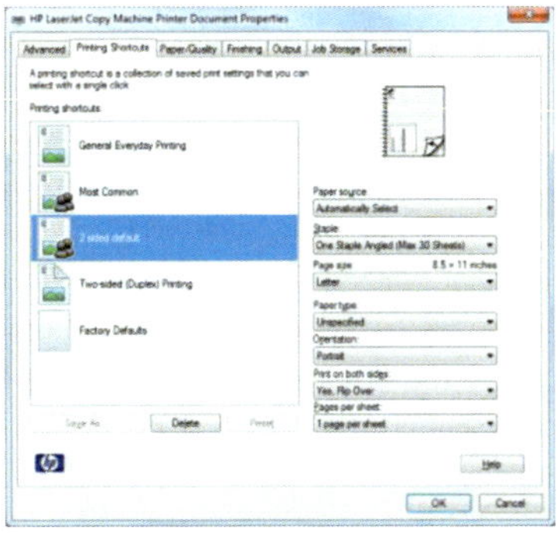

Figure 14.1 - Print dialog

A pre-structured case outline is a representation of the researcher's conceptual framework about the data that will be collected. In other words, it is an outline of the types of questions he hopes to answer during the course of the study. As the name implies, this is done before any interviews are conducted and before any data is collected. The researcher lists all the hypotheses from the usability plan, groups them into categories, and assigns short names to represent each. Of course this outline will change as the study progresses; the pre-structured case outline represents the initial framework for data analysis on which new findings and insight will be built.

Example

For example, consider a usability plan analyzing the print properties dialog box. The usability plan lists the following hypotheses:

1. The user will feel the dialog box is intimidating and complex
2. Common tasks will span multiple tabs
3. The user will not figure out how to do 4-up printing
4. The user will not figure out how to staple printed jobs
5. The label 'Job Storage' and 'Output' will not mean anything to the user
6. The user will not figure out how to create a print shortcut

The researcher notices a relationship between many of the hypotheses and some of the hypotheses may have more than one root cause. From this list, the following pre-structured case outline is created:

Variable	Tags
Efficiency	E_TAB: Time will be wasted moving between the tabs E_TIME: It will take too long to do common tasks
Learnability	L_4UP: The user will not figure out how to change the Pages per sheet control L_STAPLE: The user will not figure out how to get a job stapled L_SHORTCUT: The user will not figure out how to create a print shortcut
Familiarity	F_JOB: The label "Job Storage" will not mean anything to the user F_OUTPUT: The label "Output" will not mean anything to the user
Simplicity	S_OVERALL: The overall impression will be "complex" S_MODEL: The user will not be able to describe all the dialog does after use
Motivation	M_OVERALL: The user will not like the design
Visibility	V_TAB: The user's common tasks will contain data on more than one click

Pointers

Pre-structured case outlines are typically hierarchical. You can make the hierarchy as deep as you like and you can make the number of individual items as large as you feel is necessary, however overly large or complex hierarchies are typically more of a hindrance rather than a help. In the context of usability studies, the top-level categories are usually the variables of usability. The hypotheses are typically categorized according to the variables, though occasionally a single hypothesis could span multiple variables. Each item needs to include a label and a short description. It is usually a good idea to prefix the label with the first letter of the variable of usability it is under.

Coversheet

A coversheet, also known as a contact summary sheet, is a single sheet of paper summarizing all that was learned from a given participant. The coversheet is a living document; it contains the most current thinking and analysis pertaining to a given interview. In other words, a well-done coversheet should preclude the researcher from ever having to view the text of the interview again; the coversheet contains all the important data from the interview.

The coversheet is typically created before an interview and filled with basic information (participant number, date, demographic data, etc.). Immediately following the interview, it is a good idea to jot down a few notes on the coversheet that sum up the researcher's overall impressions. As soon as the interview is transcribed, the coversheet is affixed to the front of the document. Any subsequent analysis on the interview is reflected on the coversheet.

Example

Back to our print properties example, the finished coversheet might look something like this:

Participant 7 June 10:15am

Age: 30's Gender: F Computer experience: Novice

Overall: This user never noticed the [Properties] button on the print dialog before. There was considerable confusion as to what the dialog was for. She was even more confused when she left the dialog and was brought back to the Print dialog!

Data collected: E_TIME, L_4UP, L_STAPLE, L_SHORTCUT, S_OVERALL, S_MODEL, M_OVERALL

Quotes: "Wooah! What is this dialog for? How do I get back to the Print dialog?"

Pointers

Coversheets are very unstructured and vary a great deal according to personal taste. Some make them very verbose containing every possible detail from the interview. Some make them very terse with just the highlights. As long as you the researcher can easily find the information you are looking for, then the level of detail is good. Note that a coversheet consisting of two pages defeats the purpose because it cannot be easily scanned.

Coding

Coding is the heart of qualitative data analysis. It is the process of marking interesting parts of the transcription for subsequent collection. This is done by tagging all the text relating to the first item in your pre-structured case analysis, then doing the same for all the other items. When the process is complete, it should be possible to easily collect all quotes and notes relating to a given hypothesis.

There are many coding techniques. Probably the simplest is to simply use a red pen and circle relevant passages of the transcription. Certainly more sophisticated strategies exist, such as using XML tags or similar constructs. My personal favorite is to use the comment feature built into most word processors. Regardless of the technique employed, three essential pieces must exist:

Highlight The part of the text relevant to a given tag must be highlighted. The participant may ramble on for five minutes in response to a question, but only a sentence or two of this response may pertain to a given tag.

Label Each code should contain a label from the pre-structured case outline.

Notes While a label itself is commonly sufficient to annotate text, the case often arises when a few words are necessary to make the code understandable.

The pre-structured case outline contains the list of labels that you expect to find in the interviews. Frequently the researcher notices patterns during the interview that were not anticipated and therefore not in the pre-structured case outline. In cases like these, add a new entry to the outline and re-code the previously completed interviews. It is not uncommon to add two or three labels to the outline by the end of a study.

Example

An example of a coded transcript is presented below using Microsoft Word's comment feature. Note the tag labels are in all-caps and the notes are in normal text.

Action 10	Clicks on the [Properties] dialog	**Commented [HJ7]:** E_TIME
Response 10	Woooah! What is this dialog for?	**Commented [HJ8]:** S_OVERALL, M_OVERALL. Clearly upset
	How do I get back to the Print dialog?	**Commented [HJ9]:** S_MODEL. User appears very confused here
Question 11	Can you tell me what this dialog is meant to do?	
Answer 11	Um, no. It is really complicated	**Commented [HJ10]:** S_MODEL, S_OVERALL

Figure 14.2 - Annotations used to tag an interview transcript

Pointers

A good job coding a transcript will greatly simplify later data analysis. It will tell you what you have collected and what you have not. It is usually a good idea to make a note on the coversheet about all the labels present in a given interview to facilitate collection activities.

Vignette

A vignette is a "focused description of a series of events taken to be representative, typical, or emblematic" (Miles & Huberman, 1994, p. 81). If this sounds like a persona, it should! The only difference is that a persona is meant to represent a segment of the target population while a vignette is meant to represent a segment of your interviews.

The vignette creation process begins when the researcher observes similarities between participants or interviews. As he notices patterns, he begins to be able to predict how a given interview will turn out. The vignette is a representation of this pattern. It describes the "typical" interview for a category of participants. Though no single interview played out exactly as the vignette describes, it broadly represents a number of interviews. It is important to remember, however, that a vignette is not fiction; every point must be backed up with data from the interview.

Example

After observing five users struggle with the print properties dialog, the researcher begins to notice a few patterns:

Title: Frightened Freddie

Support: #2, #3, #5, #7

Expert:

Frightened Freddie is trying to print his history paper in the computer lab. To his surprise, he notices another printout in the bin that is double-sided. "I didn't know it could do that!" exclaims Freddie as he walks back to his computer. After searching in vain for a "double-sided" option in the Print dialog, he happens upon the [Properties] button near label displaying the name of the printer. Clicking on the button brings up the print properties dialog in all its glory. None of the controls or text in the dialog makes any sense to Freddie. He can feel his blood pressure rise as panic sets in... "What have I done?" he mutters as he quickly reaches for the red [X] at the top of the dialog. "I will never do that again" he says as he prints his paper single-sided.

Supporting Quotes:

#2 R2	"Um, I think I did something wrong. What was I supposed to click?"
#7 R3b	"Wooah! What is this dialog for? How do I get back to the Print dialog?"
#5 R3	"This is way too complicated."
#2 R6a	"Could you please just tell me were to click?"

Pointers

When all of your participants completely fit into existing vignettes, you have reached the point of saturation. This means you have gathered all the data you are likely to gather with your existing usability plan. The good news is that this means you are done; no more data collection is necessary. The bad news occurs when you have not collected data for some of your hypotheses. This might mean you need to adjust your usability plan to address these issues.

Comprehension check

Quiz 1 **Terms and Definitions**
From memory, list the four types of qualitative data analysis tools and provide a definition for each.

Quiz 2 **Pre-structured Case Outline**
From memory, list all the properties of a pre-structured case outline.

Quiz 3 **Coversheet**
From memory, list all the properties of a coversheet.

Quiz 4 **Coding**
From memory, list all the properties of coding.

Quiz 5 **Vignette**
From memory, list all the properties of a vignette.

Problems

Problem 1 **Pre-structured case outline**
Consider the following situation:

> You are a member of a small team building a web site for a high school track team (note that this is the same situation discussed in the previous chapter). You are nearing the end of the project and would like to know how the stakeholders are going to respond to the site. The manager of the team has tasked you with analyzing the collected data (from last chapter's exercises) and report on your findings. Specifically, she wants to know about any defects, overall impressions of quality, opportunities for improvement, and which parts will be most heavily utilized.

Create a pre-structured case outline for this situation.

Problem 2 **Codes**
Based on the pre-structured case outline from Problem 1 and based on the data collected from last chapter (Problems 1 through 6), code your data. Make sure you:

- Highlight the text pertaining to a given code
- Mark the text with the relevant code
- Make any additional comments as to how the text relates to the code

Problem 3 **Coversheet**

Create a coversheet for each piece of data coded in Problem 2. Make sure each coversheet includes:

- A brief (1-2 sentence) summary of the findings
- A mention of the data collected (which codes are represented in the transcript)
- Important or noteworthy quotes

When you are finished, the coversheet should be so complete that you are not tempted to read the transcript again.

Problem 4 **Vignette**

Create one or two vignettes describing the interviews. Each vignette should:

- Have a title or a name
- Reference to which participants the vignette was drawn from
- Consist of approximately one paragraph
- Have a list of supporting quotes from actual participants

Standards

There are two purposes of this unit. The first is to educate us about the current state of the art in user interface design. This includes the best practices and ideas that are in use around us. The second part is to sharpen our analysis skills as we critique established standards

15 Layout

The method by which controls are laid out in a dialog, content is arranged on a page, or information is presented on a screen has a lot to do with how the user will be influenced by and react to the design. With all aspects of HCI, as it is with communication in general, layout is part art and part science. While an understanding of layout principles may help the designer make better decisions, there will always be an element of creativity and personal judgment.

There is a large overlap in the principles of layout as they pertain to HCI and graphic design. Therefore much of what we know about HCI layout was originally described in the older and more established graphic design discipline. However, while graphic design is primarily concerned with the layout of static items on a page or screen, HCI is primarily concerned with the layout of interactive and dynamic items. Thus the usability principles of efficiency, simplicity, learnability, and mapping are much more important to the HCI designer than the graphic designer.

The design principles of layout most relevant to HCI are: grouping, contrast, balance & symmetry, unity & consistency, reading order, and rhythm.

Grouping

Grouping:

The process of encouraging the user to infer a relationship exists between design elements

Grouping is the process of encouraging the user to infer a relationship exists between design elements through the use of visual clues. When done correctly, grouping can be a powerful mapping tool because it helps the user to recognize that elements are related. Grouping can be obvious or subtle, can relate a large number of controls or as few as two. Any time the designer needs the user to associate related controls, the principles of grouping can be applied. The designer has several grouping tools at his disposal, the most common being enclosing, visual similarity, and proximity.

Enclosing

Enclosing is the most obvious grouping method because it leaves no room for doubt that a relationship exists. In other words, it is an inductive (as opposed to deductive) grouping mechanism. We see enclosing at work in virtually every user interface design around us:

Application The operating system groups the controls of one application separately than those of another by enclosing them in an application window.

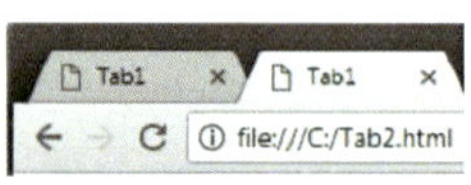

Document Browser controls are grouped differently than web page controls because the browser controls are enclosed in the browser chrome while the web page controls are enclosed in the page frame itself.

Dialog Controls are grouped in a dialog because they are enclosed in the dialog frame. Virtually all window managers have a standard way to enclose dialogs.

Figure 15.1 -Windows application
Figure 15.2 - Google Chrome
Figure 15.3 - OS-X dialog

Figure 15.4 - Toolbar

Toolbar Controls are grouped in a toolbar because they are enclosed by the toolbar itself.

The enclosing principle of grouping is at work any time a region of an interface is separated from the rest of the design. This is most commonly done with lines, boxes, and shading.

Visual Similarity

Grouping can also be achieved through visual similarities. Here, the user is asked to infer that controls are related because they have a similar size, shape, coloring, or drawing style. Because group membership must be inferred, it is called a deductive grouping mechanism. There is a seemingly endless amount of variation in the application of visual similarity grouping strategies. Some of the most common are:

Figure 15.5 - Maximize / minimize
Figure 15.6 - Cut, Copy, and Paste

- **Color**: related items are drawn in one color where non-related items avoid the color. This can only be used to set apart one or two groups; more will cause the design to appear overly cluttered and confusing.
- **Size**: a group of controls can be set apart by making them much larger or smaller than the ambient controls. As with color, this must be used sparingly. Most users will not notice subtle differences in size or will think that size has no meaning if there are too many size-delimited groups.
- **Style**: a reader of a magazine should be able to differentiate advertisements from the magazine content due to style shifts. Since the content in an article tends to use a single font, color, and spacing strategy, it is readily distinguished from advertisements using a different style.

While it is generally acceptable to set apart one or two groups with visual similarities, the technique should be used sparingly. It could result in a design feeling disjoint or noisy, discouraging notions of grouping rather than enforce it.

Proximity

A final grouping tool is proximity where the user is encouraged to believe that controls are related because they are close to each other. This, as you have probably guessed, is another deductive grouping mechanism. The rule of Proximity simply states that the closer two controls are to each other, the closer their relationship. However, the rule is a bit broader than that; controls are implied to be related if they are aligned horizontally or vertically, as well as being next to each other. Quite possibly, proximity is the most commonly used grouping tool. A small set of common examples are:

- The OK and Cancel button at the bottom of a dialog.
- The font name, font style, and font size controls in the character formatting dialog of a word processor.
- The edit control and the search button for a search engine web page

Regardless of the grouping tool that is used, one hard and fast rule must be followed: if controls are related group them; if controls are not then don't. It is not uncommon for designers to mistakenly or unintentionally group unrelated items, falsely implying that they are related. This is a great way to achieve a negative mapping score!

Contrast

Contrast:

The process of creating a pattern then breaking it

The principle of contrast can be defined as creating a pattern then breaking it. The human brain is very good at noticing patterns and attention is immediately drawn to the exception. A designer can leverage this principle to draw attention to the most important elements in the design.

Contrast can be achieved in a seemingly infinite number of ways. Regardless how it is done, the two necessary components are the pattern and the exception. Rather than attempting to enumerate all possible ways to achieve this, a few examples will be presented.

The first example is a special type of contrast called white space. White space, otherwise known as negative space, is the process of drawing attention to a part of a design where less controls exist. This is done because a pattern of density is established in the rest of the design. However, when one part of the design lacks this density, attention is drawn to it. Perhaps the most famous example of this was the debut release of the Ford Mustang in 1964. At that time, newspapers were the most common way for people to get news. Newspapers were uniformly dense with every square inch of the paper filled with articles and advertisements. Ford took out full-page newspaper ads at great expense to announce the release of their new car. Rather than fill the page with details about the car, they left almost the entire page empty with just a small picture of the car at the bottom. There was such a huge amount of contrast due to the effective use of White Space that the ad was the talk of the town for many weeks.

Second, consider a stopwatch app for a mobile device. The main screen is divided between controls and status information. However, the most important component in a stopwatch app is the start/stop button. The user should be able to find it immediately because any delay will result in a less accurate time measurement. To draw attention to the most important control, contrast is used. The designers carefully colored every control grey except one: the start/stop button. Since the pattern was set (grey controls), the eye is immediately drawn to the exception (the green button). If the entire design were colored in green, the start/stop button would be much less noticeable.

Finally, consider an eCommerce site such as Amazon. While it is important to draw the user's eye to the products, the most important item on the screen is the [Buy] button. If the user cannot readily find how to add items to her shopping cart or to checkout those items, then the design is a failure regardless of its other attributes. It thus makes sense to make the [Buy] button contrast with the rest of the site. This can be accomplished by setting it apart from the rest of the controls, using a different drawing style than the other buttons, or by using a contrasting color. In each of these cases, the user's attention will be drawn to the item the designer deems most important.

Balance

Balance:

Creating a design that appears even

Balance is the process of creating a design that appears even. Imagine a fulcrum below a dialog; which way will it tip? A balanced dialog will not appear to tip to one side or another. When balance is used in a design, the viewer is given a sense of calm and comfort; people naturally like things that are balanced. When a design is off-balance, tension exists. There are three main types of balance that can be employed: symmetrical, asymmetrical, and radial.

Symmetrical balance is perhaps the easiest to use in a design. Imagine a line drawn through the middle of a design. For a design to use symmetrical balance, every element on the right side of the line must have a corresponding element on the left side. This corresponding element does not need to be identical; it simply must have a comparable optical weight.

Asymmetrical balance is the process of balancing items of different optical weight by offsetting them different amounts from the center. The heavier an item feels, the closer it needs to be to the center. Thus a large or prominent element on one side could be balanced by a light or less prominent element on the other only if the less prominent item is much further from the centerline. Unfortunately, there is not a simple way to compute an element's optical weight. Some influencing factors include:

Color Bright or contrasting colors have greater optical weight than neutral or muted colors.

Value On a white background, dark items have greater optical weight than light colors. The opposite is true for black backgrounds of course.

Size Larger items appear heavier than smaller items.

Texture Items with more intricate details appear heavier than simple items.

Direction Elements with an obvious orientation such as a car, a human face, or an airplane tend to appear heavier when the direction is further away from the viewer. A human face will feel balanced when it is facing the camera and in the center of the frame or when it is facing to the right on the left side of the frame.

Asymmetrical balance is a powerful design tool because it can arouse emotion in the viewer. It can impart a sense of resolved tension and drama; appearing at first to be off but then coming together the longer it is viewed.

The final type of balance is also the least commonly used. Radial balance occurs when there is a center-point in the middle of the design and elements on one side of the point have a corresponding element on the other side. Here, attention is drawn to the center of the design, even if the center is lacking any design elements at all.

Unity

Unity:

The process of making all the items in a design appear to belong to a greater whole.

Unity is the process of making all the items in a design appearing to belong to a greater whole. Thus the viewer's eye will not be drawn so much to the individual elements of the design so much as the overall impact of the design as a whole. Unity can be achieved in a design if each element contributes to the same message. In a painting, each component should draw from the same palette and use a similar style. In a song, each movement should convey similar emotions using similar musical mechanisms. In a document, similar formatting should be used throughout and the text should convey a similar tone.

In the context of HCI, Unity is tightly connected with the programming concept of Cohesion. Cohesion, as you may remember, is the property of a function where everything is directed to accomplishing one task. Thus if a dialog is cohesive (doing one thing and one thing only), then it is likely to convey a sense of unity. If the main screen of a mobile app is cohesive, then it is likely to convey a sense of unity. Designers often fail in this sense when they try to cram functionality into an interface where it does not belong. This is particularly prevalent when adding enhancements to an existing design. It is often easier to simply add a few controls to an existing design rather than take the time to properly integrate the new functionality. The result of such breakdowns in unity can be simplicity problems (the required mental model grows more than necessary), mapping problems (users fail to form a valid mental model when they fail to recognize the uniqueness of extraneous functionality), and visibility problems (the more things that are added to a design, the more difficult it becomes to find the information or control the user actually needs).

Another important aspect of unity is the familiarity quality of consistency. Not only can Unity be influenced by what a design does, but also how it is done. In other words, if a design asks the user to enter text one way in part of a design and another elsewhere, then the lack of consistency can make the design feel less unified.

Usually designers run into unity challenges when they get overly focused on one component of a design and fail to think about the big picture. The fix is simple in theory but difficult in practice. In theory, one needs only to make a continual investment in understanding (or at least being familiar) with all aspects of the design, keeping the most important users and scenarios in mind, and think about the overall user experience. In practice, however, there are often marketing, managerial, and technical pressures to lose this perspective.

Reading Order

Reading Order:

The path the user's eyes follow as it moves through a design.

Reading order is the path the user's eyes follow as it moves through a design. While of course the path is ultimately up to the user, aspects of the design can lead the user's eye along a pre-determined path. There are two factors influencing reading order: the reader's natural disposition to scan a design in a given way and the use of lines and other visual clues.

Children are born with no predisposition in how to scan the world around them. Thus, the way we scan a printed document, a web page, or even the road while driving a car is learned. In order to predict the user's reading order on a given design or to create a design encouraging the user to follow a given reading order, it is necessary to understand the social norms of the user. Western readers, for example, start at the top left of a page and scan from left to right. Middle Eastern readers, on the other hand, start at the upper right corner and scan right to left. Internet users are trained to start scanning a page in the upper right corner, first scanning across the top then scanning across the left column. Users without much internet experience follow a different reading order, often focusing on the center of the screen. The best way to understand the preferred reading order of your target audience is to observe them.

Reading order can also be influenced by visual clues. The human eye tends to follow lines and arrows, shying away from dark or bland parts of a design in favor of bright or interesting parts. Artists often use subtle clues in a painting to direct the eye towards the intended focus point. Notice how lines and the gaze of people tend to be directed towards the center of a composition. The more aware you are of this graphic design tool, the more you will see it at work in advertisements and works of art around you.

An interface designer should take reading order into account in every design. This can be done by carefully identifying the sequence of actions the user will need to perform to accomplish the important scenarios, and ensure that reading order is well supported by the design. Remove distractions that may lead the user astray and ensure the reading order is as simple and direct as possible. If the sequence of actions the user needs to accomplish to perform a mainstream scenario requires the user to jump over all parts of a design, then it is likely that reading order improvements can be made.

Rhythm

Rhythm:

A strong, regular, and repeated pattern.

Visual Rhythm might just be part of unity. Rhythm is a strong, regular, and repeated pattern. As mentioned previously, the human brain is very good at finding patterns in the world around us. A design should gratify this propensity by introducing patterns as well as adding variations to the patterns for interest.

When we are exposed to a constant and repeated sound such as the hum of tires on the pavement when driving or the gentle hiss of a heater in a house, we are comforted but bored. When this pattern is altered with periods of intensity and periods of calm, especially if these periods too can be fit into a pattern, the listener is interested. This is, in fact, why we find music appealing. Music is a combination of patterns at a number of different levels, each varied in a creative yet structured way.

Consider a movie consisting of nothing but 10 second close-ups. While it may keep the viewer's interest for a few minutes, she will shortly become impatient and loose interest. Instead, the director introduces a variety of camera angles and scene lengths, keeping the viewer's interest throughout the entire movie.

Consider text consisting of nothing but five-word simple sentences. Five words is not many. At first it is fine. Soon it becomes very boring. Every sentence is the same. It is putting me to sleep. We lost the reader's interest. We lost the writer's interest! Wait! I just did something here to break the pattern. I start with a simple sentence. Next, a compound sentence that goes on much longer than one might think a sentence should, using phrases and other devices to keep the reader's attention. When we want to be clever, a complex sentence is introduced. Not too many, too few. (Clark, 2008)

Text, sequences of web pages, dialogs in a desktop app, slides in a presentation, and screens in a mobile app all need to take rhythm into account. While each individual design should be true to its purpose, it should serve as a part of larger composition.

Examples

Example 1 **Super Hero**

Figure 15.7 – Drawing of a superhero

Grouping Notice how the buildings are all drawn in the same style. Even though their shapes and sizes are very different, they all are clearly members of the group called "buildings."

Contrast The vast majority of the image is devoted to the buildings which, as was previously mentioned, are all part of a group. This pattern is broken by the figure in the foreground. Drawn in black and rendered in a different style, he stands out from the rest.

Balance The heavy figure on the left is mostly offset by the large build on the right. As a result, the balance tilts to the left, towards to the focus of the image.

Unity The entire image is created from a very small palette of colors and a small number of shapes.

Reading Order The eyes of the super hero, the light lines in the background, even the building the super hero is standing on, all these things focus the eye towards the lower center of the image.

Rhythm No two elements are the same even though they all clearly belong together.

Example 2 **Harbor**

Figure 15.8 - Painting of a harbor scene

Grouping The three structures are grouped using visual similarity. They are clearly part of the same foreground group. The same can be said about the mountains in the background and the birds.

Contrast There is a strong color pattern of sea-green and grey, broken by the roofs of the structures. This makes the structures stand out.

Balance The light of the sky offsets the heavy texture of the two large structures and the large mountain to the right. This makes the entire image appear to tilt to the right, towards the focus of the image.

Unity The entire image is created with the same graphic style and the same limited palette. This makes every element feel like it is part of a greater whole.

Reading Order Everything about the image draws the eye to the right.

Rhythm Every bird in the foreground is different, every detail of the structures are different and every boat in the background are different. Yet they are all the same somehow.

Example 3 **Color Dialog**

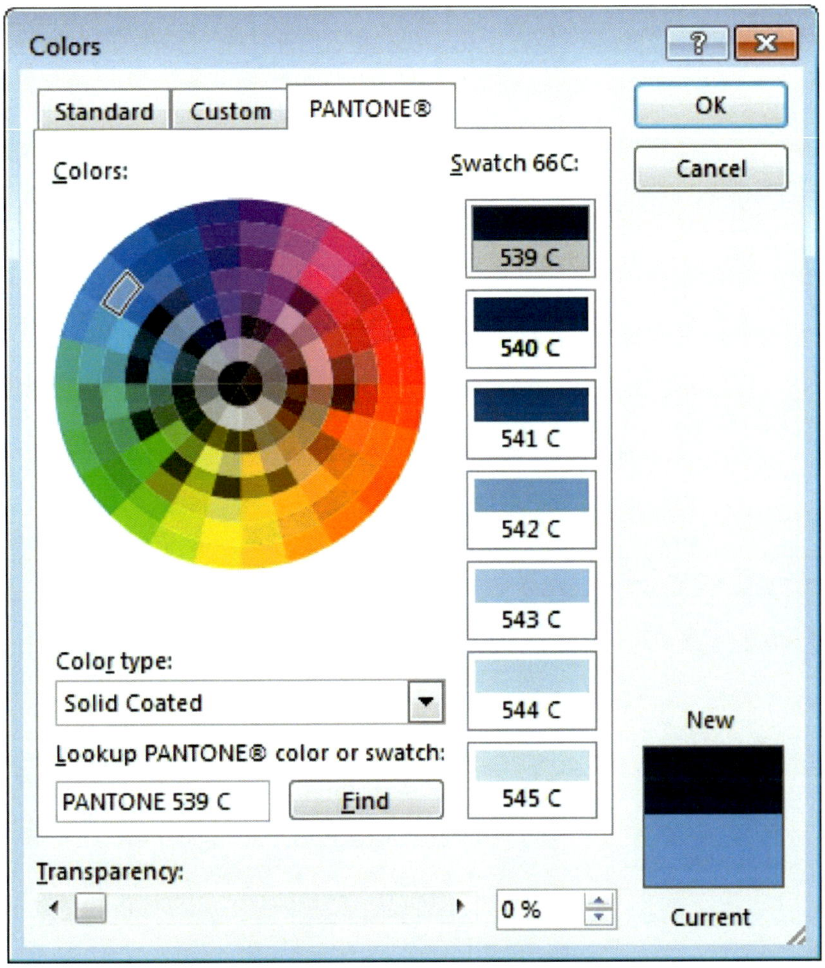

Figure 15.9 - The PANTONE® color dialog in Microsoft Publisher

Grouping The parts that are within the dialog are grouped away from the dialog chrome through bounding. The parts within the PANTONE® tab are similarly bounded within the dialog. The seven color rectangles under the "Swatch 66C" label are grouped by proximity and visual similarity.

Contrast The principle of contrast is not really demonstrated here.

Balance The heavy weight of the color wheel to the left of the dialog is somewhat offset by the OK/Cancel buttons on the right. Overall, the dialog feels too heavy on the left.

Unity The same typeface and font size is used throughout the dialog. The line width is also consistency used throughout. The selection metaphor for the color wheel is unique and breaks unity somewhat.

Reading Order There is a poorly-defined reading order here where the user is encourage to start with the color wheel, move on to the swatch, and finish with the OK/Cancel buttons. However, these controls are not arranged in a line and they leave out many of the controls from the dialog (such as "color type" and the [Find] button.

Rhythm The variety of control types adds interest to the dialog. Perhaps too much.

Example 4 Apple OS-X Finder Window

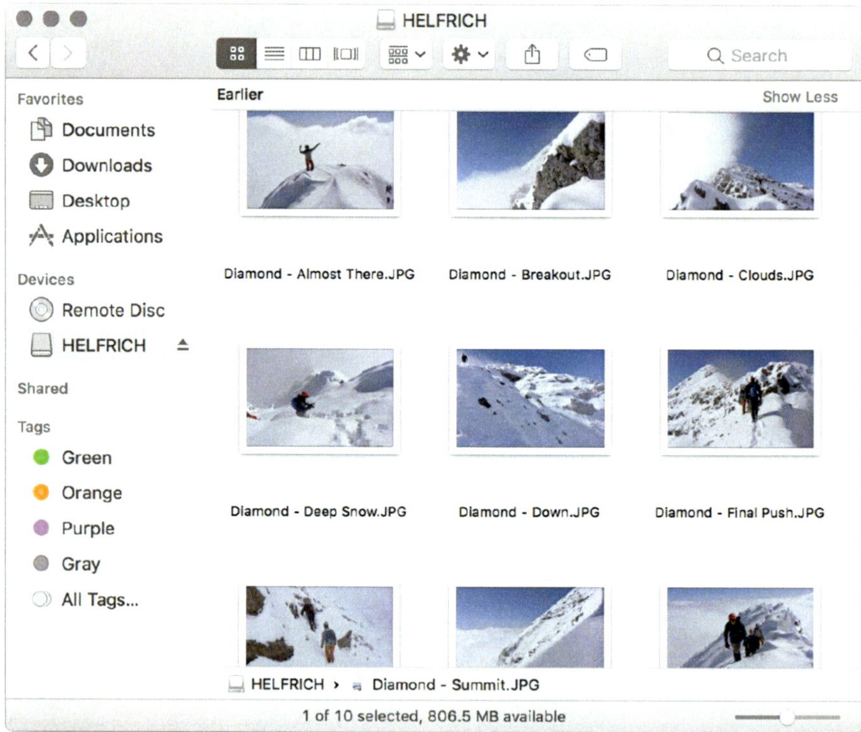

Figure 15.10 - Apple OS-X finder window

Grouping There are many enclosing mechanisms in use with all the lines and rectangles. Proximity is at play with the controls running along the top of the dialog. Finally, there are visual similarity elements at work with the size of the icons in the main window, the windowing controls in the upper left corner.

Contrast All the elements are drawn with the same style and the same color, yielding very little contrast. The difference appears when one hovers over the windowing controls in the upper left corner.

Balance The dialog is very top and left heavy. This, unfortunately, draws the eye away from the interesting parts of the dialog (the large white square).

Unity The consistent use of color and drawing style makes all elements appear part of the greater whole.

Reading Order There is no obvious scan direction through the dialog. This leaves the user with the unavoidable questions: where do I go and what do I do?

Rhythm The dialog has enough variety to provide some interest, but the dominant impression is: boring. The more important question is how this dialog fits with the whole operating system. Again, the lack of variety in interface elements and drawing style makes this window feel uninteresting.

Comprehension check

Quiz 1 **Terms and Definitions**
From memory, list the six types layout tools and provide a definition for each.

Quiz 2 **Problems and Solutions**
What layout tool can best help in the following situation: I have a unique look and feel for my game but then bring up a standard system dialog to save the preference file.

Quiz 3 **Problems and Solutions**
What layout tool can best help in the following situation: While watching members of the target audience use my e-commerce site, I realize that their mouse needs to cross the entire screen three times to make a single purchase.

Quiz 4 **Problems and Solutions**
What layout tool can best help in the following situation: All the controls on my mobile app are on one side of the screen, leading the app to feel too empty on the other.

Quiz 5 **Problems and Solutions**
What layout tool can best help in the following situation: On my blog, some readers cannot tell the difference between the content and the ads.

Quiz 6 **Problems and Solutions**
What layout tool can best help in the following situation: The most important button on my mobile game is the "Fire" button. However, some users don't even realize it is there!

Quiz 7 **Problems and Solutions**
What layout tool can best help in the following situation: All the pages on my web site look exactly the same. I want to introduce some variety without any of the pages clashing.

Problems

Problem 1 **Milky Way**

Critique the following picture according to the principles of graphic design layout.

Figure 15.11 - Painting of man looking up to the Milky Way

Specifically describe how this image achieves each of the following and critique how well it performs:

- Grouping
- Contrast
- Balance
- Unity
- Reading order
- Rhythm.

Problem 2 Rowing a Boat

Critique the following picture according to the principles of graphic design layout.

Figure 15.12 - Painting of a man rowing a boat into the night

Specifically describe how this image achieves each of the following and critique how well it performs:

- Grouping
- Contrast
- Balance
- Unity
- Reading order
- Rhythm.

Problem 3 **WordPad**

Critique the following desktop application according to the principles of graphic design layout.

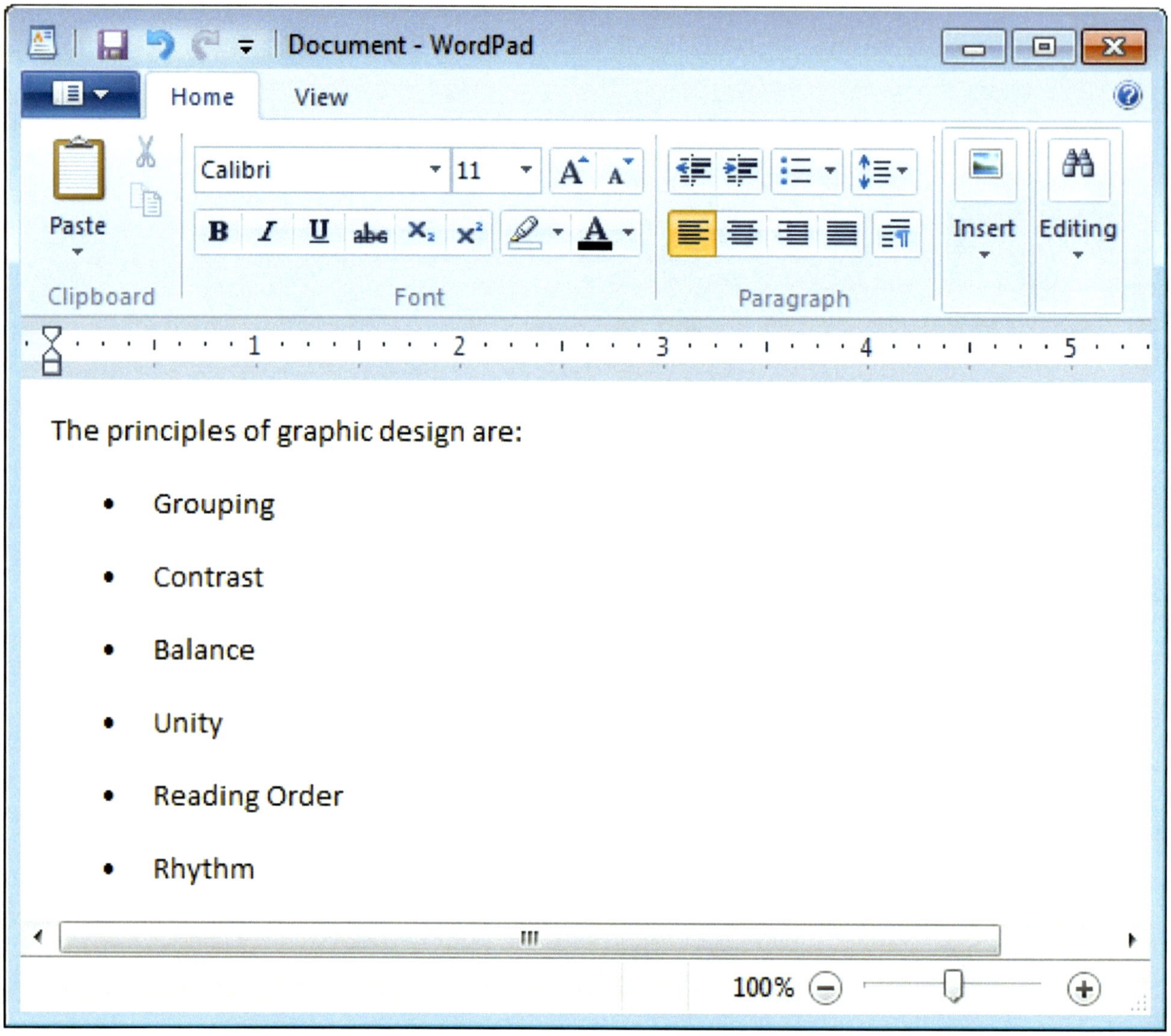

Figure 15.13 - Picture of Microsoft WordPad

Problem 4 **Paragraph Dialog**

Critique the following dialog according to the principles of graphic design layout.

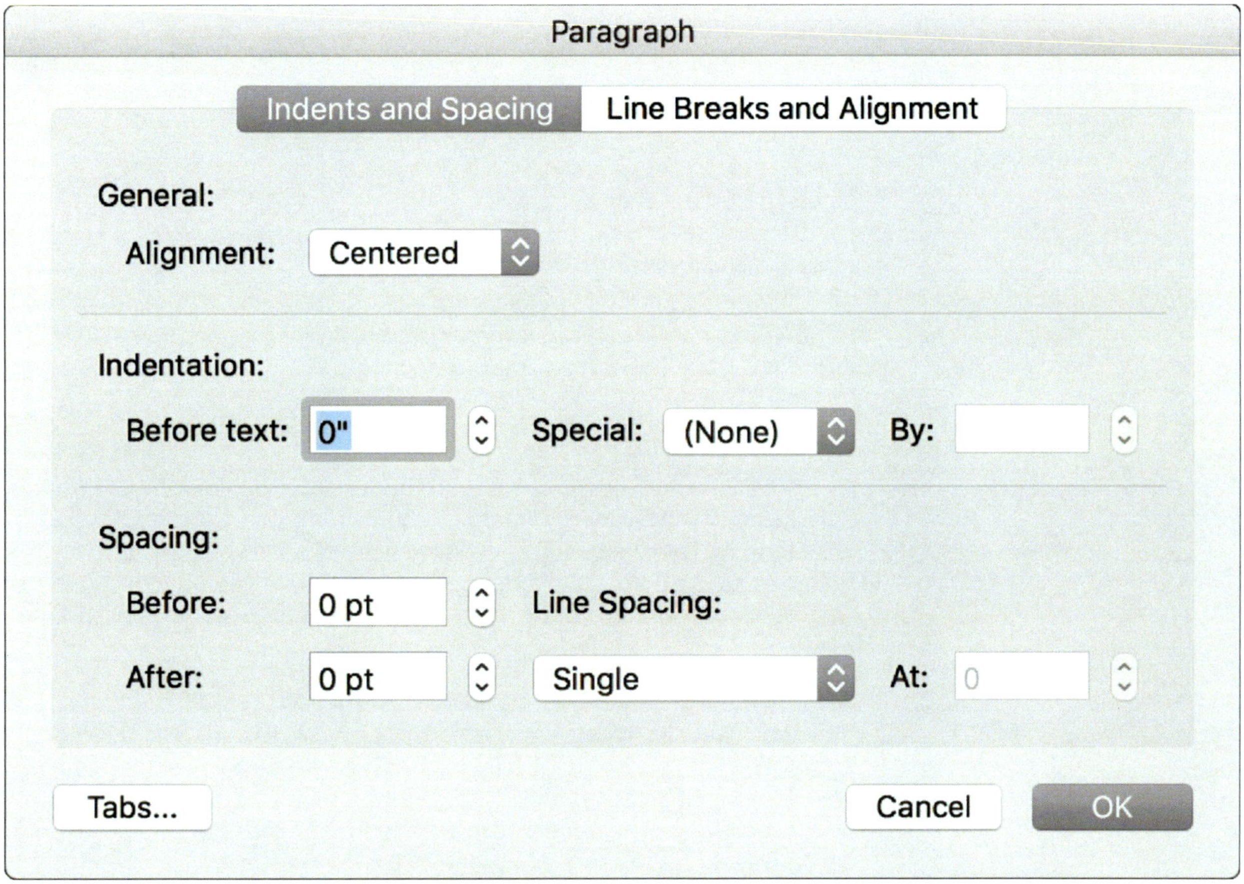

Figure 15.14 - Paragraph dialog on Microsoft Word for Macintosh

Problem 5 **Web Application**

Find a web page to a site that you visit frequently. Take a screen-shot of the page and critique it according to the principles of design layout. Suggest design alterations that could overcome the design's shortcomings.

Problem 6 **Mobile Application**

Consider a mobile application that you use frequently. Take a screen-shot of a screen and critique it according to the principles of design layout. Suggest design alterations that could overcome the design's shortcomings

Problem 7 **Desktop Application**

Consider a desktop application that you use frequently. Take a screen-shot of a screen and critique it according to the principles of design layout. Suggest design alterations that could overcome the design's shortcomings

16 Color

There are two aspects of color that pertain to HCI: the psychology of color and color theory. The first pertains to how colors convey meaning and emotion to people, the second to how color is perceived. Both must be taken into account when choosing colors for a given design problem.

Color Theory

Color Theory:

The science of mixing colors so they appear pleasing to the eye.

Color theory is the science of mixing colors so they appear pleasing to the eye. To understand why two colors look good together it is necessary to understand how the brain interprets color, how the eye perceives color, and how light itself carries color information.

Light

Our notation of color, of course, originates from light. Light itself has no color. Light is simply electromagnetic (EM) radiation traveling at a certain wavelength. This wavelength varies greatly from the length of a football field to smaller than the diameter of an atom. We can only perceive light that has a wavelength of about a nanometer; the rest is ignored by our eye.

Most of the light we encounter in our daily lives is mixed, containing many different wavelengths. Light emitted from the sun, for example, is composed of photons of virtually all wavelengths. Some of these wavelengths are filtered out by the atmosphere and thus do not make it to the earth. It is relatively rare to encounter light consisting of only a single wavelength, lasers being one of the rare exceptions.

Since this light contains information that is highly useful to navigate our world, it is not surprising that we were given the ability to perceive light. Our eyes perceive light at the wavelengths where enough useful information is present to make it possible to negotiate our world and the things contained therein.

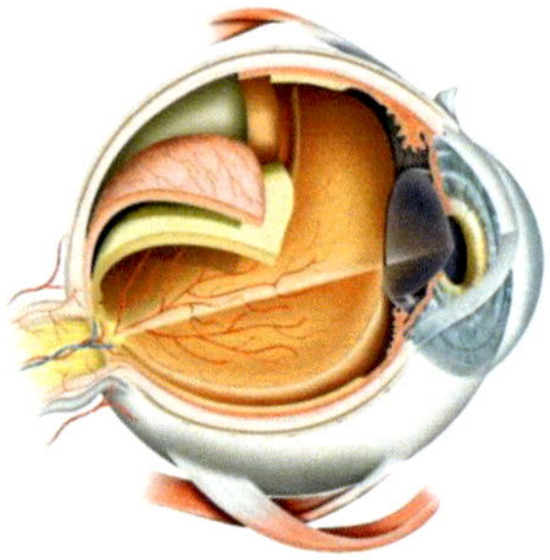

Figure 16.1 - Figure of the anatomy of a human eyeball

Biology

We sense light through our eyes using four types of photoreceptors: rods which sense light from a wide variety of wavelengths, red cones sensing light from about 500 - 700 nm, green cones sensing light from 450 - 650 nm, and blue cones sensing light from 400 - 500 nm. Our perception of color originates from these cones. If, for example, our red cones perceived light from 100 - 200 nm, then apples would no longer be red to us. Thus color is not a property of the physical world; instead it is an artifact of how we perceive the physical world. In other words, color is a human abstraction used to understand the information present in the light around us.

Everyone does not have cones tuned to the same optic frequency. Some individuals have overlapping photoreceptor ranges for their red and green cones, making it impossible to differentiate these colors. We call this condition "color blind." Note how the photoreceptor range of most people overlap quite a bit. This means that people, in general, are better able to differentiate red from blue or green from blue than they are green from red.

Color Spaces

Because we have three types of photoreceptors in our retina, our eyes send three dimensional color data to our visual cortex (the part of the brain that processes visual information). Thus we receive data in the RGB color space.

A color-space is a mechanism for representing color. Possibly the easiest way to do this is to simply name colors (such as red, hot pink, and brown). While this may be convenient, it does not provide any insight into the relationship between colors (such as pink is halfway between white and red) nor does it provide a way to describe all possible colors.

Figure 16.2 - The RGB color space

The **RGB** color space designates color according to the amount of Red, Green, and Blue light mixed to describe a given color. We call the RGB color space an "additive color space" because adding equal parts of Red, Green, and Blue light yields white. This is a convenient color space because it is close to how the retina perceives light (assuming that the definition of Red, Green, and Blue are closely aligned with the middle of the receptive range of red, green, and blue cones). It is also convenient because it is possible to directly convert RGB values into color with a device emitting appropriate amounts of Red, Green, and Blue light. Computer displays are one such device. However, the wavelengths of the light emitting from the Red, Green, and Blue diodes of most computer displays does not directly map to the Red, Green, and Blue photoreceptors of the human retina. As a result, colors may appear differently on different computer displays if they are not calibrated correctly.

Figure 16.3 - The CMY colors space

Most of the color we perceive in our daily lives does not originate from a light emitting display. Instead, it comes from light shining off a surface. Consider a white light shining on a blue sheet of paper. We perceive the surface as blue because only blue light enters our retina. However, the paper is anything but blue. It is, in fact, everything but blue! The white light, consisting of all possible wavelengths, hits the blue paper. The paper absorbs all the wavelengths of light except blue that is then reflected back to the viewer. Thus adding colors to the paper will not produce white as it does in the RGB color space, but rather produce black; the more colors that are added, the more light that is absorbed. We call this a subtractive color space, the color space of pigments. The base colors of the pigment color space are the opposites of the base colors in the light color space: Cyan, Magenta, and Yellow. Cyan opposes Red, representing the inverse of the photoreceptive range of the human Red cone. The same is true of Magenta and Green as well as Yellow and Blue. This color space is called CMY (Cyan Magenta Yellow) or, when black is added to the mix, **CMYK** (K standing for "Key").

Both the RGB and the CMY color-space are sufficient for describing all possible colors. However, they do not provide any insight into what colors go together. To address this need, it is first necessary to define what it

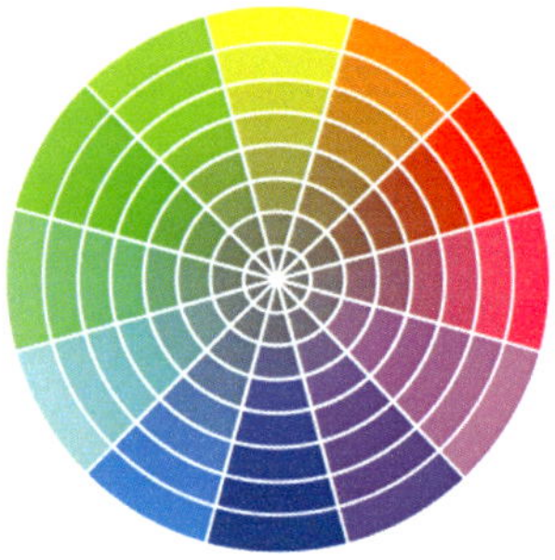

Figure 16.4 - HSL color space

means for colors to "go together." The human brain is very good at spotting patterns and, in fact, finds patterns pleasing. For example, patterns are what differentiate noise from music. We appreciate music because our brains find the patterns in music satisfying. In much the same way, the human brain looks for patterns in light and, if they are found satisfying, we like the color combination. The visual cortex (again, the part of the brain that processes visual information) converts color values from the retina (red, green, and blue) into three different variables:

Hue The average wavelength of the perceived color. We don't represent hue as a wavelength in nm, instead as an angle from 0 degrees to 360 degrees.

Saturation How pure the color is. Saturation is measured as a percentage where 0% means grey and 100% means the color is as bright as it can be.

Luminosity How bright the color is. Luminosity is also measured as a percentage where 0% means black and 100% means white.

The **HSL** (Hue, Saturation, Luminosity) color space consists of these three values. We will be using this color space exclusively when describing different strategies for mixing colors.

Figure 16.5 - LAB color space

One final color-space is necessary. This color-space originates from a type of processing the visual cortex performs on RGB input. Possibly this is best described by example. Consider a man looking for an orange. Because the orange does not emit light, the way we perceive its color is influenced by the light shining on it. Up to this point, we have always described pigments according to how they react to white light. However, how would this same orange appear in the twilight where the ambient light is almost red? How would it appear in the middle of day where all light filters through a canopy of green leaves? In each of these cases, the orange would appear a different color (in the RGB color-space) because different color light is bouncing off it. To compensate for this, the human visual cortex has the ability to discern the color of things taking into account the dominant color of ambient light. You can see this effect in action by wearing a pair of colored glasses for a few minutes. Notice how your eye compensates for the color of the glasses and after a while you can see the color of things normally. Now take the glasses off. For a few seconds, the sky and the trees look weird. This is because it takes about a minute for your visual cortex to adjust to a new ambient light color. The **LAB** or **CIE** color space was designed to represent color accurately taking into account this color shifting. It was developed for photographers who must pay special attention to ambient light color or there will be a tint or cast to skin tones.

Color-Space Conversions

As mentioned previously, light is composed by many wavelengths of light. The human eye converts this multi-dimensional (one dimension for every wavelength) data into three-dimensional data through the three cones. Cameras have different receptive ranges than the retina and thus different RBG values are derived. In other words, the conversion of natural light into RGB values depends on the receptive ranges of the photo-sensors.

Computing CMYK values from RGB requires definitions for Cyan, Magenta, and Yellow. A true Cyan, the standard used for RGB → CMYK conversions, would be the inverse of Red. That being said, the CMYK colors in a printer commonly have a different composition.

Most designers use a software tool to convert between the various color-spaces. The standard Microsoft Windows and Apple Macintosh color dialogs do this transformation.

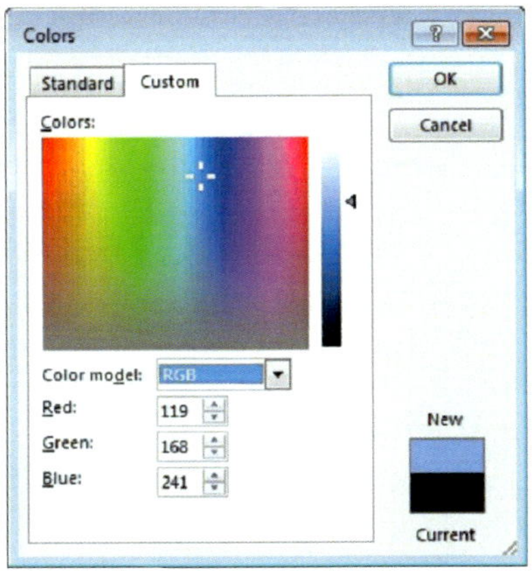

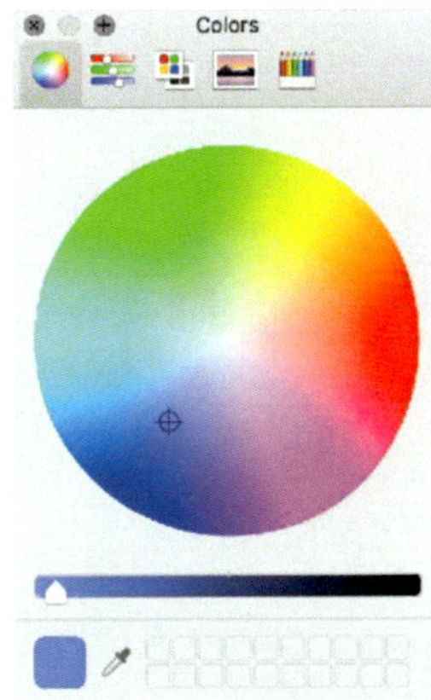

Figure 16.6 - Microsoft Windows color selection dialog
Figure 16.7 - Apple Macintosh color selection dialog

Mixing Colors

It is often necessary to find collections of colors that appear appealing to humans. To predict how people will react to color combinations, it is necessary to take into account how humans sense color (with red, green, and blue photoreceptors) and how humans process color in the visual cortex.

Figure 16.8 - Tints and Shades

The simplest way to mix colors is called **Tints** and **Shades**. A tint is variation of a color where white was added. The result of adding white is to increase luminosity and slightly decrease saturation. A shade, by contrast, is achieved by adding black. This serves to decrease luminosity and decrease saturation. It is always safe to create a variation of a color using a tint or a shade; the resulting color will match the base color. Finally, a **Tone** is a color whose saturation has been reduced without altering the hue or luminosity, serving to make a color look dull.

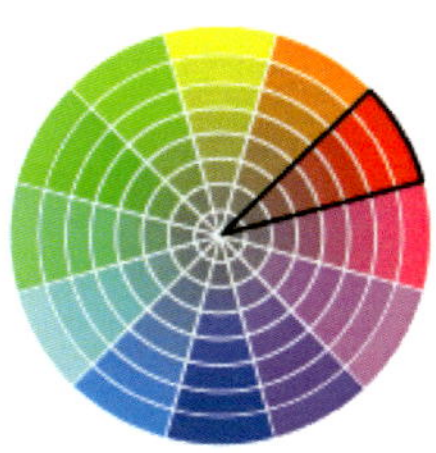

Figure 16.9 - Monochromatic colors

Monochromatic colors are collections of colors with a constant hue but different degrees of saturation and luminosity. This is usually achieved by using tints and shades. Monochromatic colors are easy to manage; any combination of saturation and luminosity tends to look good together. Monochromatic colors are also balanced, summoning feelings of uniformity and connection. However, monochromatic colors lack any real contrast and, as a result, do not feel vibrant.

Figure 16.10 - Analogous colors

A slightly more complex way to mix colors is to make small variations in hue while using larger variations of saturation and luminosity. This color scheme is called **Analogous**. Generally, analogous color schemes are visually appealing but tend to lack contrast or interest. If the change in hue is kept small, the colors will appear harmonious. However, if the change in hue is too great, the color scheme will fall apart and appear random.

Figure 16.11 - Complementary colors

The **Complementary** color scheme is achieved by mixing colors of opposite hue. This is the color scheme of maximal contrast. As a general rule, it is a good idea to make one of the two colors dominant, using the base color 80% of the time and the complementary color 20% of the time. Complementary colors often look noisy and are difficult to manage. Use them carefully, only when a strong point or maximal attention need to be raised. To compute a complementary color, it is first necessary to find the HSL from the RGB. Next, add 180° to the hue. Finally, convert the HSL back to RGB.

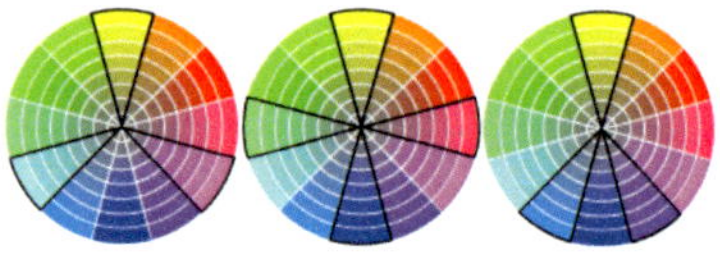

Figure 16.12 – Triadic, tetradic, and split complemenary colors

Other more complex color schemes include **Triadic** (colors separated by 120° on the color wheel), **Tetradic** (a combination of two complementary colors where the hue differences are small), and **Split Complementary** (a mixture of complementary and analogous). In general, complex color schemes should be used with caution. If they are not used correctly, the colors feel random.

Color Psychology

Figure 16.13 - Cabin demonstrating cool and warm colors

Aside from picking colors that look good together (Color Theory), another important consideration when choosing colors is the psychological implications of colors on most people. This can be best explained with an example. Imagine a cold night in the mountains. A winter storm has just blown in and several feet of fresh snow is on the ground. You were cross-country skiing for a couple hours and are making your way to a cabin. As you round a bend, the cabin comes into view. The cool blue of the night is broken by the amber glow of the cabin lights. There is something about the colors themselves that spark an emotional response. The indigo blue of the sky and the mountains feel cold but the amber lights of the cabin feel warm.

Color psychology is the process of identifying and using people's emotional response to colors to convey meaning. People respond to colors. This is a fact that we cannot change. We can either use this reaction to our advantage in an effort to use colors to further our message or we can let our color choices unintentionally drive the user's mood.

For the most part, a person's emotional reaction to color is independent of their social-economic background. Yellow always conveys energy for example. However, some cultures have special associations with certain colors. White is used for weddings in Western countries, for example, while red commonly adorns a bride in China.

Generally, the more saturated the color, the more intense the emotional response to the color. The designer can always dial down the effect of a color by reducing saturation. For example, brown is a shade of orange (de-saturated and lower luminosity). It has the same psychological implications as orange except muted. A bright orange room would be too much for a restaurant, but a room colored beige, tan, and brown would be appetizing.

Red is the most emotionally intense color, associated with such raw feelings as anger, violence, ambition, sex, and energy. Use the color red carefully because there is always an emotional response. A red tie, a red car, a red flag will always draw attention. This is why stop signs are red, warning messages are red, and provocative lipstick is red.

Orange is the color of youthful energy, adventure, rejuvenation, and social interaction. Orange spurs people to action, not in the raw way red does but in a positive and enthusiastic way. Orange is a great choice for a restaurant or a gathering spot because it encourages people to linger and be active

There are two sides to **yellow**: intellectual energy and positivity. Yellow is the color of confidence and logic, stimulating thinking more than the other colors. Yellow is also the color of optimism, enthusiasm, and fun. Overuse of yellow can give the impression of lack of seriousness and bring fatigue.

Green is a soothing and harmonious color, surfacing such emotions as balance, compassion, and nurturing tendencies. When people are stressed, they turn to green. When people are exhausted, green appeals to them. When people are over stimulated by their surroundings or their life, they can be often found seeking after green things.

Blue is the color of stability, authority, and trust. It is a calming color, not likely to rouse people to action but rather to convey a sense of stability and satisfaction. Authority figures wear blue. Religious institutions often incorporate blue. Institutions trying to convey a sense of loyalty and responsibility often turn to blue. Blue is a cold color.

Purple and **violet** are the colors of respect, power, royalty, wisdom, and independence. It is considerably more edgy and energetic than blue, but much less so than red. Purple conjures feelings of spirituality, not in the sense of religion but in the sense of independent or personal spirituality: prayer, meditation, curiosity, and mysticism.

White is the color of purity, used to convey a sense of cleanliness, innocence, and being pristine. White is also simple, stark, empty, and often boring. White conveys a sense of goodness and godliness and with it a sense of perfection. It can also be used to convey a sense of efficiency being the most uncluttered color.

Black is the color of mystery and power. It brings a sense of secrecy and mystery, that there is more than meets the eye. Black is a color of intimidation, often bringing feelings of fear. People wear black when they want respect. Black is the color of formality, being both conservative and serious at the same time.

Observe that adjacent colors on the color wheel have similar emotional responses. Half the colors convey energy and action to some capacity while the other half convey stability and inaction. We call red, orange, and yellow **warm colors** because they tend to brighten the mood and draw attention to themselves. We call green, blue, and purple **cool colors** because they make people more somber and reflective. Most pictures with both warm and cool colors tend to have the cool colors fade into the background while the warm colors vie for our attention.

Examples

Example 1 **Super Hero**

Analyze the following image according to color theory and psychology.

Figure 16.14 - Drawing of a superhero

Color Theory The sky has two colors: (51, 84,126) and (72, 108, 147). These correspond to hue 213° so they are tints of each other.

The buildings are (88, 58, 84), (125, 96, 134), (157, 167, 205), and (132, 137, 177). These correspond to the hues 307°, 285°, 227°, and 233° respectively. Thus they are analogous.

Finally, the super hero is black (0, 0, 0) and purple (132, 137, 177). This, it turns out, is the same color as the bright part of the building.

Altogether, the color scheme of this image is analogous.

Color Psychology These colors are all cool. Blue is associated with stability and trust while purple is associated with respect, power, and independence. All of these meanings contribute to the feelings the author of this image are portraying.

Example 2 **Harbor**

Analyze the following image according to color theory and psychology.

Figure 16.15 - Painting of a harbor scene

Color Theory The sky has many colors, ranging from: (147 178, 160), (97, 123, 110) and (225, 250, 190). These correspond to hue 150°, 145°, and 155°. These correspond to hue 150°, 145°, and 155°. This is a very narrow range of hues. Also, we can see that our saturation is between 15% and 20%, meaning the color does not have much punch. As a result, they can be considered analogous.

The buildings are (226, 90, 32), (222, 88, 59), and (93, 41, 20). These correspond to the hues 18°, 11°, 17° respectively. They are analogous (almost monochrome!) with respect to each other. Note that they are opposite sides of the color wheel from the sky (off by about 20°). Thus the overall scheme of the image is split complementary (a mixture of analogous and complementary).

Color Psychology The sea-green that dominates the image is a mixture of green (soothing, balance, harmonious) and blue (stability, trust, and calming). These are in sharp contrast to the red of the roof. Note that the red is about 90% saturated meaning that the emotional intensity of the color is high. Red brings to the surface feelings of anger, ambition, and energy. It is unclear why the author of the image chose those emotions in an otherwise peaceful scene.

Example 3 **Wedding Photographer**

Analyze the following image according to color theory and psychology.

Figure 16.16 - Image of a wedding photography home page

Color Theory The colors of the dress (225, 221, 220), drapes (221, 216, 212), and chair (200, 201, 195) have nearly identical saturation of 2%, 4%, 3% respectively. The bride's skin (204, 157, 136), hair (66, 41, 45), and the wood floor (135, 124, 120) have similar hues (18°, 350°, 16° respectively), but the saturation is so low (33%, 38%, and 53% respectively) that they appear very similar. The title is the same color as the model's hair and the rest of the text are a shade of wedding dress (166, 146, 143). The overall color scheme is thus analogous.

Color Psychology The dominant color is white with the page background, the dominant color of the picture, and the bride's dress all being whites. This brings the feelings of purity, cleanliness, innocence, and goodness. Since these are the feelings that most brides would wish to convey with their wedding pictures, the colors are well-chosen. Note that the page title, the bride's skin, and the bride's hair are contrasting with this color scheme, bring appropriate interest and attention to the focus of the page.

Example 4 **Weather Desktop Application**

Analyze the following application according to color theory and psychology.

Figure 16.17 - Local weather from the Microsoft Windows 8 weather application

Color Theory The sky ranges from (3, 4, 41) to (0, 2, 99) to (27, 115, 185). These correspond to the hues of 238°, 239°, and 207° which are all analogous colors representing tints (the clouds) and shades (the dark gradient at the bottom) of blue. The sun is (251, 233, 108) which is 52°, almost exactly 180° from the blue. Thus the overall color scheme is split complementary.

Color Psychology Though there is a bit of yellow in the sun icons, the vast majority of the pixels are blue. This is a very cool color, giving the feelings of low-energy and peace. Furthermore, the blue color is often associated with stability and satisfaction. Note that the exact scene was meant to represent the current weather, but the extreme use of a very saturated blue (ranging from 66% saturation all the way up to 100%) seems to be intentional. After all, most cloudy skies are not that blue! One could say that the emotional response to the color selection is heavy-handed and over-kill for the needs of the application.

Comprehension check

Quiz 1 **Color Psychology**

From memory, list the typical emotional responses to the following colors:

- Red
- Orange
- Yellow
- Green
- Blue
- Purple
- White
- Black

Quiz 2 **Terms & Definitions**

From memory, define the following terms:

- Cool color
- Warm Color
- Tint
- Shade
- Tone
- Hue
- Saturation
- Luminosity

Quiz 3 **Color Spaces**

From memory, define each of the following color spaces:

- RGB
- CMYK
- HSL
- CIE

Quiz 4 **Color Schemes**

From memory, define each of the following color spaces:

- Monochromatic
- Analogous
- Complementary
- Split-Complementary
- Triadic
- Tetradic

Problems

Problem 1 **Milky Way**

Analyze the following image according to color theory and psychology.

Figure 16.18 - Painting of man looking up to the Milky Way

Problem 2 **Rowboat**

Analyze the following image according to color theory and psychology.

Figure 16.19 - Painting of a man rowing a boat into the night

Problem 3 **Weekly Calendar Mobile Application**

Analyze the following mobile application according to color theory and psychology.

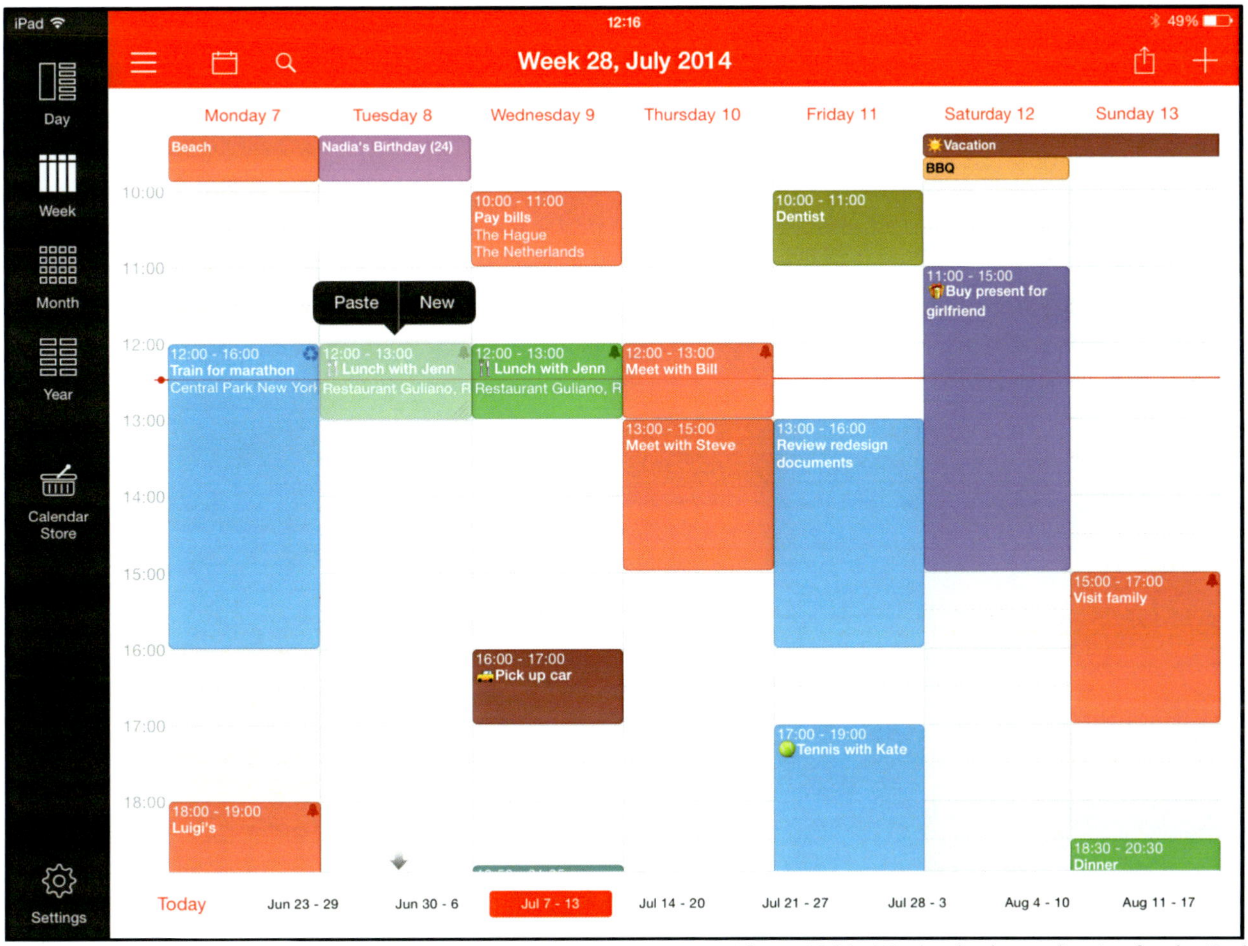

Figure 16.20 - Week calendar application for the iPad

Problem 4 **Microsoft Office**

The toolbar design of Microsoft Word 2007 is the following:

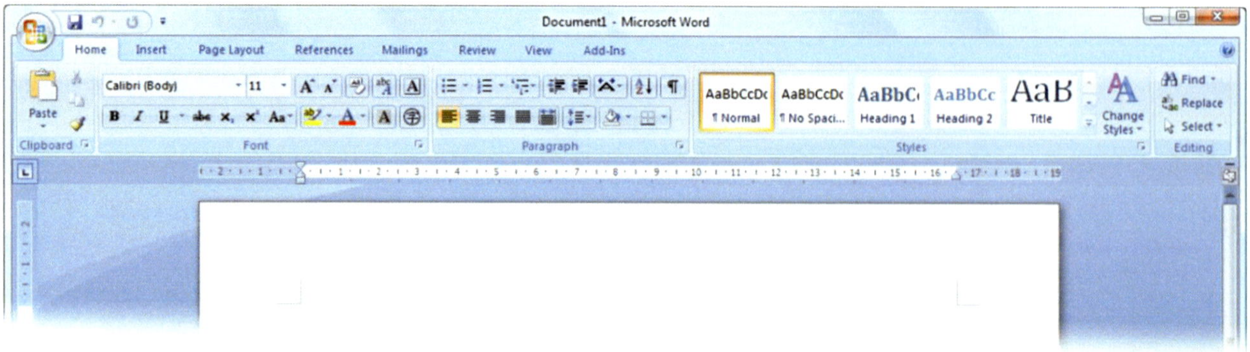

Figure 16.21 - Microsoft Word 2007

The following is the same view of Microsoft Word 2013:

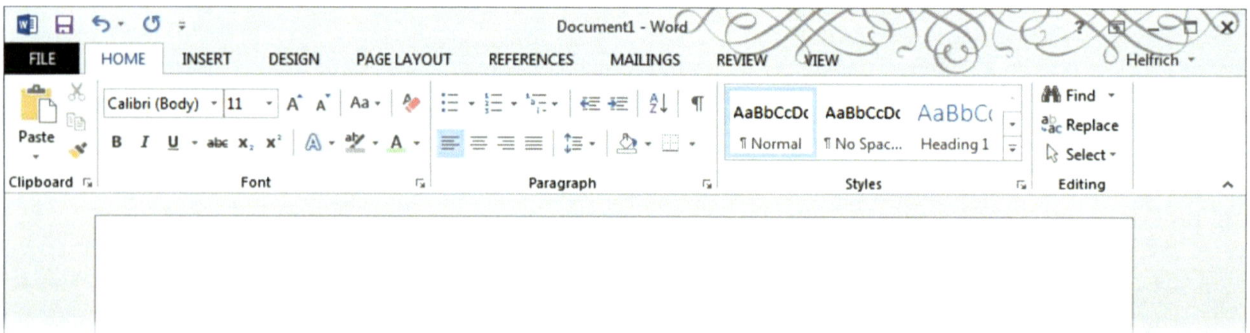

Figure 16.22 - Microsoft Word 2013

Analyze both designs according to color theory and psychology. Which design is superior?

17 Typography

Typography is the process of conveying meaning through the appearance of text. Possibly Robert Bringhurst put it best: “Typography exists to honor content.” (Bringhurst, 2002) This is quite different than the process of conveying meaning through the text itself. In this case, the appearance of the text on the page or screen is the communication medium.

Typography is on display everywhere in the world around us. We see its influence in the layout of your textbook, design of web pages, and layout of a dialog. Typography heavily influences the impression and message of your resume. Probably the most prominent and impactful use of typography in our everyday lives comes from advertisements, company logos, and similar marketing material. Every time text is used in a document or a user interface, typography is sending the user a message. The designer needs to ensure that this message is consistent with the message that needs to be sent.

Properties of a Typeface

Perhaps the best way to discuss the properties of a typeface is through a bit of vocabulary. The first such term is a **stroke**. A stroke is a single mark of a letterform. This can be thought of as a gesture or stroke of the pen when drawing the letterform. The letter ‘O,’ for example has one stroke. The letter ‘T’ has two, and the letter ‘Z’ has three. The most common strokes used in the Latin alphabet include:

Spur: The ending slash of a capitol ‘G’. Note that originally a spur was an accent, being a special form of the letter ‘C’

Arc of a stem: The curve at the end of a ‘j’. In many languages, the arc of a stem remains an accent, such as: ¿ ţ ç

Bar: The horizontal line in the middle of a ‘t’ . If the ends of the bar are not free, such as an ‘A’, it is often called a crossbar. If only one end is free such as an ‘E’, it is often called an arm. A bar can also be called a cross stroke. The bar too was originally an accent, which it remains in many Eastern European languages: ī Ŧ Ħ

Bowl: The enclosed oval or circle in a ‘b’. If the bowl is not completely round, such as an ‘e’, it is called an eye

Counter: an open bowl such as ‘c’ and ‘u’. This is also commonly called an aperture or an open counter

Apex: the point where two strokes meet. Often this juncture is embellished with a decorative marking. An apex when the juncture is at the bottom of the glyph is called a crotch

Figure 17.1-6 - Parts of a glyph

j i ? ;

Dot: The point resting below a question mark '?' or above a 'j.' Technically speaking, these are accents. Eastern European languages have both the dotted 'i' and the dotless 'ı' called an iota. Other names for the dot include the jot and tittle: ċ

d 1 f 6

Ascender: The part of a glyph extending above the midline: 'd'. Ascenders provide important visual clues as to the identity of a letter, being much more prominent than other strokes

j p q y

Descender: The part of a glyph extending below the baseline: 'y'. As with ascenders, fonts with prominent descenders are easier to read because the human eye is able to more easily recognize the shape of words

Figure 17.7-9 - Parts of a glyph

g g

Figure 17.10 - Two glyphs for the same letter

A **glyph** is a single manifestation of a letter, comprised of one or more strokes. There can be many glyphs associated with a single letter. For example: look at the glyphs on the left to see two different ways to draw the letter g. In many ways, one can think of a typeface as a collection of glyphs.

A **letter** is a symbol corresponding to a spoken sound. A letter has semantic meaning. For example, twenty glyphs representing the letter 'a' are still the letter 'a' in the English language. In most cases, a given typeface will have exactly one glyph for each letter. However, some typefaces give the designer more options.

ﬁ fi

Figure 17.11 - FI as a ligature and not as a ligature

A **ligature** is the combination of two or more overlapping glyphs into a single glyph done for the purpose of increasing readability. In most cases, the human reader will not realize that a ligature is used. The most common ligatures used are 'f' and 'i' into ﬁ and 'f' and 'l' combined into ﬂ. High quality fonts have dozens of ligature pairs while inexpensive fonts commonly have none. While ligatures may seem like more trouble than they are worth, Guttenburg's original font had more than two hundred of them, each individually carved in metal by hand.

æ ae

Figure 17.12 - Diphthong and the two letters it was created from

A **diphthong** is a combination of letters used to symbolize a linked sound. There are few examples in the English language, but they are quite common in Eastern Europe. Examples of a diphthong is a + e → æ.

Typeface Styles

Typeface designers have quite a bit of latitude in designing glyphs while still making them true to their letters. The main tools they have at their disposal include serifs, stress, slant, weight, decorations, and style.

Serif:

A serif is a small line or embellishment at the end of a letterform designed to increase readability.

Some typefaces include little feet at the end of bars and arms. These marks are called **serifs** and are used primarily to increase readability on the printed page. As the eye moves over large amounts of small text, serifs provide subtle clues as to the identification of glyphs making it easier for the reader to recognize the letters. Serifs actually decrease readability on low resolution displays as are commonly found on many computer screens. If the DPI (dots per inch) are less than 200 and the text is rendered in less than 15 points, then serifs should be avoided. Serifs can be used when a more decorative look is desired for the text while sans serif (typefaces lacking serifs) convey feelings of simplicity and modernity.

Stress:

The ratio between the thinnest and thickest part of a bowl on a letterform

Another property of a typeface is **stress** or the difference or ratio between the thinnest part of a bowl and the thickest. Some fonts have zero stress (or 1:1 ratio), meaning that bowls have uniform thickness throughout their shape. Most serif fonts have a 1:2 ratio, called a moderate stress, to increase readability. Generally, fonts with a pronounced stress (ratio of 4:1 or greater) are considered more elegant and decorative than those with zero stress.

Slant:

The angle of stress in a glyph

Slant is the angle of stress in a glyph. It is usually measured by the line passing through the thinnest parts of a bowl such as 'O'. Vertical slant is called "Transitional." When the angle is roughly 15 degrees, the natural angle of holding a fountain pen, the slant is called "Humanist". Fonts with a Transitional stress are often considered boxy and orderly.

The thickness of a font is called the weight. Weight has a large impact on the perception of heaviness of the text on the page. There are many degrees of weight in a typeface:

Typography

Hairline: The thinnest possible rendering of a font. Bars are typically rendered in a single pixel, regardless of the resolution of the printer or the screen. While Hairline fonts convey a sense of elegance, they can be difficult to read

Typography

Lightface: About half the thickness of a typical font, Lightface conveys much of the meaning of Hairline without the readability problems. Lightface is a good choice for titles

Typography

Book: The standard weight for a typeface. It is loosely defined as the weight commonly found in large runs of printed text such as a book or newspaper. Book is optimized for readability

Typography

Medium: Slightly heavier than Book, Medium can be applied to a heading or a title to great effect. Medium is not obviously different than Book (in which the body text typically set) but does draw a small amount of attention to itself

Typography

Demi: Demi is Latin for "half" meaning halfway between Book and Bold. Because it is not obviously a different weight than Book or Bold, it should never be mixed with either in a single run of text. Instead it can be used to set off a heading

Typography

Bold: The Bold typeface is obviously thicker than Book and can be mixed into Book to emphasize or draw attention to small passages. It is difficult to read large passages of Bold text

Typography

Extra Bold and **Ultra Bold**: These two variants of Bold are used purely for decorative purposes. Their weight makes them difficult to read, making them inappropriate for use in large runs of text

Typography

Black: The heaviness weight of a given font, Black should be reserved for decorative applications such as logos and titles

Figure 17.13-20 - Weights of typefaces

The size of a typeface on the screen is measured by the point size where there are 72 points to an inch. For most fonts, 12pt is the minimal readable size when viewed on a screen. For paper applications, 10pt is often acceptable. However, it must be noted that this minimum size varies widely with the typeface. The important properties influencing the optimal size of a typeface include the message the typeface needs to convey, the feature size of the glyphs, the reading surface (quality of paper, screen resolution, etc.), and the x-height. Of these, perhaps only the x-height demands further explanation. Consider four imaginary horizontal lines running through a run of text:

- **Base line**: the line that the text appears to rest on. Note that this line is defined visually, not geometrically; glyphs commonly do not rest exactly on the baseline.
- **Ascender line**: This line roughly corresponds to the top of the highest point of a glyph in a given typeface. The cap-line may be the same as the ascender line or, in some typefaces, the ascender line may be slightly higher.
- **Descender line**: This lowest point in a glyph, this line, combined with the ascender line, defines the vertical height of the typeface.
- **x-Height**. The x-height is the highest point of lowercase letters lacking ascenders. This is typically defined with the lowercase 'x', hence the name.

Figure 17.21 - Height measurements of a typeface

Typefaces with a low x-height (where the height of the text is dominated by ascenders and descenders) can often be easier to read than those with a high x-height because the human eye picks out the ascenders and descenders to recognize words. For this reason, it is often very difficult to read blocks of ALL-CAPS or SMALL-CAP text. However, when the point size is small, there is less space to render the distinguishing features of a glyph. For this reason, text with a high x-height are often considered easier to read when the text is small.

Italic typefaces were invented in the 1800's to more closely resemble cursive handwriting. Italic was not originally a property of a font but rather a separate font altogether. Eventually italic fonts got paired with standard Roman (non-italic) font to become a font-family. In the truest form, italic is not a slanted version of its Roman counterpart, but rather a softer and more flowing version. It is difficult to read large blocks of italic text; italic is best used to emphasize or draw attention to short passages of text.

Mixing typefaces

Many applications require more than one typeface: one for the bulk of the text and a companion to serve as a contrast. This is commonly necessary for headings or captions in text, user interface controls in web pages, and chrome in desktop or mobile applications. There are no hard-and-fast rules for mixing typefaces. There are, however, a few general guidelines.

The first guideline is to make differences obvious. Mixing 11pt and 12pt fonts looks like an accident whereas mixing 11pt and 16pt looks intentional. Mixing dark grey text which charcoal looks like an accident whereas mixing dark grey with cranberry looks intentional. Recall that contrast is the process of establishing a pattern then breaking it. In this case, the variation from the base text needs to be an obvious departure.

The second guideline is to pick one or two variables to change, not to change everything. A couple common variables to alter include:

- **Weight**: Book contrasted with demi work well together. So too do book with hairline.
- **Style**: Mix a serif font with a san-serif where the letterforms, stress, slant, and x-height are similar.
- **Color**: Mix a black text with another color. As a general rule, text on white backgrounds should be dark and text on black backgrounds should be light. Text without a large color contrast tends to be unreadable.
- **Size**: Text should be 33% different in size to contrast well. The larger the difference, the more decorative the text appears. It is not uncommon for there to be a 10x difference between body and heading size.

Line Layout

When laying out a large block of text, such as an essay, resume, or web page, it is necessary to take into account line layout (the spacing of characters next to each other on a single line), paragraph layout (multiple lines of text), and page layout (columns and margins). These three levels should work in harmony to convey the intended meaning of the designer.

Line layout is primarily concerned with the density of text on a line and the amount of space between letterforms. In the days of mechanical typefaces, each glyph was embedded on a rectangular metal stamp called a letter-punch. Since letter-punches are packed next to each other, spacing between glyphs was constant. With the advent of digital typefaces, it became possible to achieve arbitrary spacing between glyphs.

VA

Figure 17.22 - Example of kerning

The first component of line layout is kerning. Kerning is the process of adjusting the spacing of adjacent glyphs (called "kerning pairs") so their spacing appears consistent and uniform. This is especially apparent with 'A' and 'V'. If the spacing is determined by their bounding rectangular rectangle, they would appear too far apart. When they are kerned, they overlap and appear more uniformly spaced. When fonts are created with little effort or expense, kerning is ignored and the spacing between letterforms appears random. High quality fonts have extensive kerning tables where many kerning pairs are carefully defined to ensure the font looks pleasing on the printed page.

The second component of line layout is character spacing, also known as tracking. While all fonts have a native tracking between glyphs, it often becomes necessary to specify non-default character spacing to achieve a given effect.

Typography

Loose: Extra white-space between glyphs, ranging from "a bit more than normal" to several ems (where an em is the width of a capitol "M"). Loose spacing is common on logos and titles used to convey a sense of airiness and elegance

Typography

Normal: This is the default character spacing for a given font as specified by the "kerning table," a table within the font describing how much space exists between each glyph pair. Normal character spacing is optimal for readability

Typography

Tight: Any character spacing less than normal yet still easily readable. It is often used in select runs of text to achieve full-justification (where both the left and right side of the paragraph are straight) to avoid hyphenating too many words for line breaks

Typography

Very Tight: Very Tight character spacing appears cramped and extremely heavy. It is commonly used for headings when the point size is very large or there is extra space between the words. The result is a sense of density without too much effect on readability

Typography

Kissing: The spacing between characters has been reduced so that the glyphs are almost touching. Kissing is decorative and should not be used except for small runs of text. Usually kissing spacing is paired with extra-bold or ultra-bold weight, or fonts that have an extremely large stress

Typography

Touching: Touching spacing occurs when the glyphs physically run into each other. Aside from logos, titles, and other special purpose applications, touching spacing should almost never be used in large runs of text due to the extreme readability issue it presents

Typography

Figure 17.23-29 - Character spacing

Negative: Occurs when characters overlap to any degree. Readability is severely impacted by Negative character spacing except when the glyphs are rendered with different colors. As with kissing and touching, negative character spacing is almost exclusively paired with very heavy weight such as ultra-bold or black

0123456789

0123456789

Figure 17.30 - Old-style and modern numerals

Numbers represent a special challenge to line layout. There are two styles of numerals: old-style and modern. **Old-style numerals** have roughly the same number of ascenders and descenders and standard text. This makes the numerals blend in with the ambient font. **Modern numerals**, on the other hand, have a uniform height. They can be thought of as "uppercase numerals." It is appropriate to use Modern numerals when the numerals need to "jump from the page," when there are many numerals in the document, or when numerals appear on a line of their own. In all other cases, Old-style are a better choice.

The final component of line layout becomes important when more than one typeface, font size, or glyph appears on the same line. Here, the most important component is that the text must appear harmonious to the reader. This is usually achieved by ensuring that all the fonts share the same baseline, though matching x-height also contribute to the sense of order.

Paragraph Layout

Paragraph layout deals with multiple lines. The designer must be aware of how the reader's eye flows over the text as well as how the text appears as a block on the page. The primary considerations for paragraph layout include line breaks, line spacing, and paragraph separation.

Line breaks signify to the reader that text resumes on the adjacent line, requiring the reader to scan to the next line to continue reading text. This is an interruption in the reading process. If the line is too short, the reader will spend too much time and effort scanning for the break. If the line is too long, then the reader will have trouble finding the start of the next line. The following guidelines govern line length:

Situation	Guideline in number of characters
Single column page	Between 45 and 75 characters. Any more and it will be unnecessarily difficult for the reader to find the next line. Any less and the reader will be spending too much time jumping between lines.
"Ideal" line	For most people, the ideal line consists of exactly 66 characters.
Multiple-column	Pages with multiple columns can have narrower lines than single-column designs. Typically we should keep this between 40 and 50 characters. Any more and the page itself will be too dense. Any less and the dangers of rivers exist.
Isolated patches of text	When small, isolated patches of text are used (such as captions and pull-quotes), we can use as little as 12 to 15 characters per line of text. Here efficiency considerations are less important because we are working with small amounts of text.

The four major justification schemes are left, right, centered, and full. Left is common for left-to-right reading languages because it is easy for the reader to find the beginning of the next line once the end of the current line is reached. Right-to-left languages such as Arabic find left justification to be awkward. Right justification is thus used mostly for decorative purposes or for a small number of lines. Center justification should similarly be avoided in large blocks of text yet can serve to introduce a balance along an imaginary central line. Full justification combines the readability of left justification with the balance of center. In order to fully justify text, it is necessary to adjust the character spacing. When the paragraph is narrow, this can lead to extremely loose spacing on one line and tight on another. When short lines of text are fully justified, an optical effect called rivers can exist. Consider a column of text where, on average, only three words appears on each line. When this text is fully justified, a large amount of character spacing will be required to make the edges line up. Since the breaks between words on successive lines will be roughly aligned and the spacing is large, white lines will appear in the text. These lines are called rivers. Rivers should be avoided because they serve to distract the reader from the intent of the page.

Line spacing is the amount of white space existing between successive lines of text. Each typeface has a natural line spacing, being a function of the line height (distance between the ascenders and the descenders) plus a fixed amount of white space. The designer may choose to use a non-default line spacing to achieve a variety of effects:

Double Double space text uses an extreme amount of line spacing, taking an excessive amount of real estate and appearing optically disjointed. Generally double spacing should be avoided. Its usefulness is limited to proofing, providing the proofer ample room to hand-write comments in draft text

Wide Wide spacing is anything between double spacing (2x) and loose spacing (1.15x). When coupled with long lines (greater than 100 characters per line), wide spacing can convey a sense of airiness in the text. Generally, wide spacing is less readable than normal spacing and should be avoided for large runs of text.

Loose Loose spacing is defined as 1.15 times the normal line height for a typeface. Recently, loose spacing has become popular and several word processors default to 1.15x spacing for body text.

Normal Normal spacing is optimized for large blocks of text and should be used in most instances where more than a few dozen lines of text are required.

Tight Tight spacing includes any spacing between .95 and touching. The tighter the spacing, the greater the impression of density of the text. Headings and titles should use tight spacing to strengthen the connection between the lines

Touching This occurs when glyphs on successive lines touch or overlap. Touching spacing should only be used for decorative purposes such as logos

Paragraph separation is the process of introducing formatting to separate paragraphs. This can be accomplished through a variety of ways:

- **First-line indent**: Where the first line is indented by a half inch or so. This is the most efficient paragraph separation technique because it only a small amount of real estate is used.
- **Hanging indent**: Where all lines but the first are indented a half an inch. This is decorative and has the appearance a bulleted list on the page. Generally hanging indents should be restricted to short paragraphs and small runs of text.
- **Paragraph spacing**: In a word-processor, paragraph spacing is called "space-before" and "space-after." Paragraph spacing, especially when coupled with full justification, yields strong alignment cues on the page because nothing breaks the left and right vertical lines. At a minimum, paragraph spacing should be 1.5x the line spacing or it will be too subtle

for the reader to notice. Large paragraph spacing reduces impressions of weight or density on the page.

- **Borders, shading, and lines**: Paragraphs can be set apart with boxes, either outlines (borders) or filled (shading). This should not be used for large blocks of text by can be used effectively as a decorative tool or when grouping user interface elements. However, these tools are commonly used for headings and titles.

Page Layout

Page layout describes how text appears on a page or screen. This includes margins, columns, placement of illustrations, headers, and footers. There is a large overlap in page layout techniques with those of the general principles of graphic design. In other words, balance, contrast, proximity, variety, and white-space are important elements to take into account. The designer must regard the entire page as a whole and carefully consider how each element influences the overall message of the page. There are three broad strategies for page layout that the designer may wish to consider when laying out text: margins, columns, and grids.

Margins

Margin layout schemes begin with an imaginary rectangle on the page, usually an inch or so from the edge of the page. From the left side of the page (for left-to-right languages), there are also one or two vertical lines extending the edge of the page. All text, images, and other items should align to these imaginary lines. Therefore, first-line indents, hanging indents, bulleted lists, and tables should be aligned. Center align in a margin layout scheme should be avoided unless to introduce contrast.

Columns

Column layout scheme can be thought of as dual margins. Due to the reduced line length, a smaller font and narrower page margins must be used. Though it is common to make all the columns the same, there is no rule for that. However, if one column is to be different, it is usually necessary to make it very different. For example, one column could be used for text and another for illustrations.

Grids

Grids are the process of sub-dividing the screen or page into same-sized boxes. Elements are then aligned to these boxes or groups of boxes. Grid layout schemes allow for much greater layout variety than column or margin schemes while still maintaining an overall impression or order and symmetry. Generally, the larger the number of boxes, the more opportunity for variation and the more difficulty in maintaining order.

A résumé is a great example of a document which would benefit from grid layout. The following resume has a 5x10 grid with no padding between the cells.

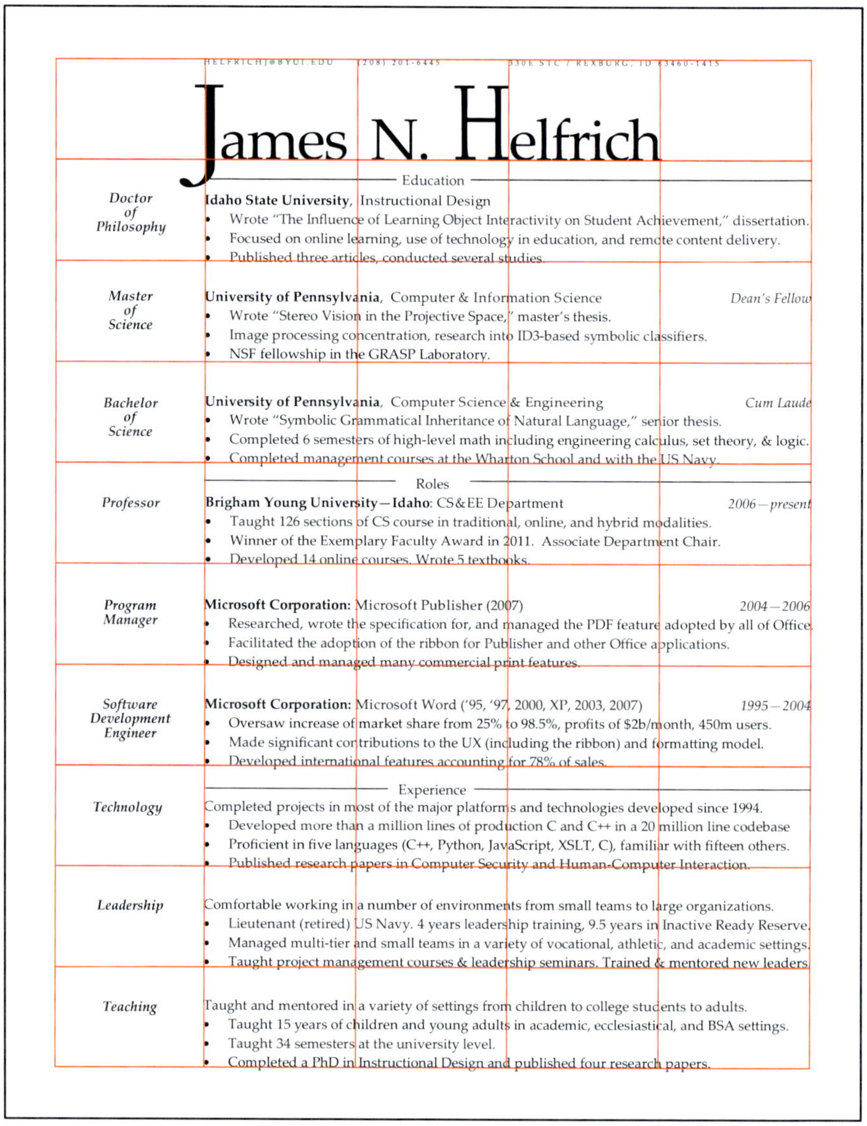

James N. Helfrich

Education

Doctor of Philosophy	**Idaho State University**, Instructional Design • Wrote "The Influence of Learning Object Interactivity on Student Achievement," dissertation. • Focused on online learning, use of technology in education, and remote content delivery. • Published three articles, conducted several studies.
Master of Science	**University of Pennsylvania**, Computer & Information Science *Dean's Fellow* • Wrote "Stereo Vision in the Projective Space," master's thesis. • Image processing concentration, research into ID3-based symbolic classifiers. • NSF fellowship in the GRASP Laboratory.
Bachelor of Science	**University of Pennsylvania**, Computer Science & Engineering *Cum Laude* • Wrote "Symbolic Grammatical Inheritance of Natural Language," senior thesis. • Completed 6 semesters of high-level math including engineering calculus, set theory, & logic. • Completed management courses at the Wharton School and with the US Navy.

Roles

Professor	**Brigham Young University—Idaho**: CS&EE Department *2006—present* • Taught 126 sections of CS course in traditional, online, and hybrid modalities. • Winner of the Exemplary Faculty Award in 2011. Associate Department Chair. • Developed 14 online courses. Wrote 5 textbooks.
Program Manager	**Microsoft Corporation:** Microsoft Publisher (2007) *2004—2006* • Researched, wrote the specification for, and managed the PDF feature adopted by all of Office. • Facilitated the adoption of the ribbon for Publisher and other Office applications. • Designed and managed many commercial print features.
Software Development Engineer	**Microsoft Corporation:** Microsoft Word ('95, '97, 2000, XP, 2003, 2007) *1995—2004* • Oversaw increase of market share from 25% to 98.5%, profits of $2b/month, 450m users. • Made significant contributions to the UX (including the ribbon) and formatting model. • Developed international features accounting for 78% of sales.

Experience

Technology	Completed projects in most of the major platforms and technologies developed since 1994. • Developed more than a million lines of production C and C++ in a 20 million line codebase • Proficient in five languages (C++, Python, JavaScript, XSLT, C), familiar with fifteen others. • Published research papers in Computer Security and Human-Computer Interaction.
Leadership	Comfortable working in a number of environments from small teams to large organizations. • Lieutenant (retired) US Navy. 4 years leadership training, 9.5 years in Inactive Ready Reserve. • Managed multi-tier and small teams in a variety of vocational, athletic, and academic settings. • Taught project management courses & leadership seminars. Trained & mentored new leaders.
Teaching	Taught and mentored in a variety of settings from children to college students to adults. • Taught 15 years of children and young adults in academic, ecclesiastical, and BSA settings. • Taught 34 semesters at the university level. • Completed a PhD in Instructional Design and published four research papers.

Figure 17.31 - A résumé rendered in a 5x10 grid layout

A few things to notice. First, if text is centered, it is centered within a cell. Second, many design elements span cells. If an element spans a cell, it takes the entirety of the cell rather than just a portion of it. Finally, the vertical location of text within the cell (this text is bottom-aligned, for example) is less important than each cell being filled in the same way. In other words, the viewer of the document does not get to see the grids as they are drawn above.

Typography and UX Design

While the applicability of typographic principles to documents or web page is clear and obvious, how do these principles apply to the design of a dialog? The answer is simple: any application that has text needs to take into account typographical principles. Specifically, one needs to take into account typographic style, line layout, paragraph layout, and page layout.

Typographic style

In most mobile or desktop application design scenarios, the choice of typeface is made by the authors of the operating system. Using a font other than that of the standard will cause the application to appear unprofessional, awkward, and untrustworthy. The exception to this rule is icon design where readability and other concerns become important. Generally, sans serif and large x-height fonts render better in these circumstances. Always use modern numerals.

Line layout

Line layout principles come into play with surprising regularity in dialog design scenarios. Be consciously aware of honoring baselines; controls should be vertically aligned not based on the bounding rectangle but rather on the text they contain. Character spacing, without exception, should always be normal.

Paragraph layout

The graphic design principle of grouping is extremely important in dialog and app layout. Paragraph layout tools such as line spacing and justification can play an important role here. In other words, line spacing should be minimal but paragraph spacing should be large so the label or title for a control can be associated as closely as possible with each other.

Group boxes are paragraph layout tools intended to encapsulate related controls by bounding them in a rectangle. The result is unambiguous grouping. Group boxes facilitate two dimensional dialog layout necessary in some complex design problems. However, group boxes introduce a degree of visual clutter and can be intimidating to some users.

Group lines are similar to group boxes except they are formed with a single horizontal line through a title or heading element. They are useful only in single column layouts. This limitation is a two-edged sword. On one hand, complex two-dimensional layouts are not possible thereby limiting the layout options for the designer. On the other hand, discouraging designers from overly complex or cluttered layouts often yields cleaner and more elegant designs.

Page layout

Page layout techniques in document layout have a large carryover into dialog and app design because they address essentially the same problem: placing interface items on the screen. As a general rule, the two layout schemes are margin and grid layout. Margin layout should be used in simple interfaces where there are less than a dozen or so design elements, grid should be used in more complex scenarios. In both cases, it is prudent to sketch the design on paper with the guidelines and distances fully specified before building the interface.

Examples

Example 1 **Currency**

Analyze the following image according to the following: typeface choice, line layout, paragraph layout, and page layout.

Figure 17.32 - a hundred dollar bill

Typeface Choice There are more than a dozen typefaces on this bill. Four are black (the "100"s, "The United States of America," "one hundred dollars", and "federal reserve note"). Most are serifs though the serial numbers and other minor characters are not. There does not appear to be much connection between the various typefaces though most are neoclassical with extreme stroke contrast and transitional stress. The aim appears to be difficulty in counterfeit rather than any single style.

Line Layout The numerals are modern. The character spacing is kissing on just about all the decorative text. This contributes to the sense of being heavily embellished. The only exceptions are the "legal tender" notification which is tight and the serial number which is monospace and loose.

Paragraph Layout There are no true paragraphs in this document. The line spacing is almost uniformly tight when multiple lines exist. This matches the character spacing.

Page Layout There is no apparent margins, columns, or grid. The document appears to be off-center, unlike previous versions of the note.

Example 2 **Meme**

Analyze the following meme according to the following: typeface choice, line layout, paragraph layout, and page layout.

Figure 17.33 – John Muir in The Mountain Trail and its Message

Typeface Choice The typeface is Eras Light ITC which is a san-serif font with decorative letterforms. The weight is light but most would call it a hairline. There is a hint of stress, especially along the vertical axis. More noticeably, the font has a slant. One would not quite call it italic. Overall, it contributes to the sense of lightness and somberness that the message and the image are trying to convey.

Line Layout The character spacing is tight but not very tight, with the characters much closer than normal but certainly far from kissing. This gives a certain amount of density to the text.

Paragraph Layout Line spacing is loose, approximately 1.25. This, combined with the very light letterforms, contribute to the sense of levity. Each line is 66 characters in length which is exactly the ideal line length. The end result is text which is easy to read.

Page Layout There is no apparent columns or grid. The text is positioned in the "white space of the image far from the busy foreground. The saunterer appears to be looking up to the text. The end result is a balanced page.

Example 3 **Scrapbook application**

The following picture is from a tablet scrapbook application. The application chrome (navigation controls, etc.) have been removed in this picture. Analyze the application according to the following: typeface choice, line layout, paragraph layout, and page layout.

Figure 17.34 - Scrapbook application

Typeface Choice The serif font is an appropriate choice given the antique look and feel of the layout. The weight is book, optimized for readability.

Line Layout The character spacing is normal, once again optimized for readability. Perhaps the title should be loose to add a decorative element.

Paragraph Layout Line spacing is tight, perhaps 80% of the normal height. This adds quite a bit of density and offsets the generous margins in the textboxes. There are 50 characters per line, a bit long for multi-column layout but certainly very readable. The fully-justified alignment emphasizes the square textboxes in which the text resides. The overall feel is compact, tidy, and readable.

Page Layout The space between the edge of the page and the picture, the indents of the textboxes, the margins within the textboxes, and the space between the textboxes are all the same. This grid (21x21 cells) contributes to a sense of order, unity, and tidiness in the overall layout.

Comprehension check

Quiz 1 **Parts of a Letterform**

Identify the part of the letter-form common to each of the following examples:

- g j p q y
- A f Ħ ī t Ŧ
- j į ţ ç
- ċ ? i j .
- b d f h k l t
- c e g h m n s u
- b d g o p q

Quiz 2 **Kerning Level**

Order the kerning level from smallest (bottom) to largest (top)

- Kissing
- Loose
- Negative
- Normal
- Touching
- Very Tight
- Tight

Quiz 3 **Typographic Units**

For each of the following definitions, identify it is a diphthong, stroke, glyph, ligature, or letter

- The combination of glyphs done for the purpose of increasing readability (fi)
- A symbol corresponding to a spoken sound
- One manifestation of a letter
- The combination of letters to symbolize a linked vowel sound (æ)
- A single line or mark of a letterform

Quiz 4 **Font Weight**

Order the levels of font Weight from darkest (top) to lightest:

- Bold
- Book
- Ultra bold
- Demi
- Black
- Extra bold
- Hairline
- Medium
- Lightface

Quiz 5 **Line Length**

What is the length of an "ideal line" for single-column work?

Problems

Problem 1

Figure 17.35 - Travel mobile application

Travel mobile application

A travel agent has produced an Android mobile application to share a trip itinerary with the clients. This application presents a brief summary of each day with a representative picture. Please critique this design according to the principles of typography. Pay special attention to:

- Typeface choice
- Line layout
- Paragraph layout
- Page layout

Suggest design alterations that could overcome some of the current design's shortcomings.

Problem 2

Figure 17-36 - Taco truck menu

Taco Truck Menu

The Taco Truck™ sells premium tacos in a large metropolitan area. A significant amount of their business is young professionals on their lunch break. These people self-identify as "foodies." On the left is their menu.

Critique the menu of the Taco Truck™ according to the principles of typography. Pay special attention to:

- Typeface choice
- Line layout
- Paragraph layout
- Page layout

Suggest design alterations that could overcome some of the current design's shortcomings.

Problem 3 **Wedding Photographer**

Critique the menu of the homepage to the "Helfrich Photography" web site according to the principles of typography. Pay special attention to:

- Typeface choice
- Line layout
- Paragraph layout
- Page layout

Suggest design alterations that could overcome some of the current design's shortcomings.

Figure 17.37 - Home page for Helfrich Photography

18 Web Applications

Web applications are defined as web sites with the functionality of an application though it is readily apparent to the user that the content is delivered through a web browser. There are several parts to this definition. The first is that the content is delivered from an external source called a web server. Web applications do not work in a disconnected environment. The second is that the web site behaves like an application, containing a variety of controls through which the user is able to interact with the content. The final component is that the user must always know that the content comes through the browser. Therefore the defining characteristic of a web application is not the technology used to deliver the content, nor is it the type of user experience, but rather the act of interacting with content through a web browser.

Though the internet in various forms traces its roots to the 1970's, the web experience as we know it arrived with the widespread adoption of HTML and the development of the first web browser. Initially web content was static in nature, allowing users to navigate between web pages but not exhibiting the rich interactivity associated with applications. This changed with the introduction of eCommerce sites in the mid 1995 such as Amazon.

Attributes of Web Computing

There are several attributes that set web applications apart from their Mobile and Desktop counterparts. These are the role of content, absence of clear application boundaries, external navigation controls, and importance of meeting the needs of novice or visiting users.

Role of content

Web applications are often dominated by content, meaning that little of the screen real-estate is devoted to interactive elements. The reason for this is partially historical; HTML was originally designed as a streamlined format for representing research and technical papers. Another reason is practical. Web content is delivered from servers over the internet, meaning it is the most convenient way to deliver up-to-date content to users around the world. People turn to the web for information and this information tends to be represented in a static, content-heavy format.

Designing for Web Applications is tricky because interactive and non-interactive elements commonly coexist on the same page. The designer needs to not only integrate the two, but also clearly advertise what actions can be performed on each item.

Absence of clear application boundaries

Browsing the web is a fluid experience; there is no demarcation between web sites. As a result, it is easy for users to lose track of where they are.

This is in stark contrast with Desktop applications where the operating system clearly demarks programs and each window is labeled.

Branding:

Visual clues in a product or user interface designed to inform the user what he is using and where he is.

The ramifications of this attribute are far reaching. The designer needs to establish a unique look and feel to a web site or web application so the user knows where he is when he arrives and when he is about to leave. This unique look is called **Branding**.

External navigation controls

In Mobile and Desktop applications, the designer is completely in control of how a user reaches a given interface element and where he can go next. This is not true with web applications. External navigation controls such as the address bar and the forward/back buttons enable the user to view pages or screens in any order. In other words, there is no guarantee that the user has to enter through the front door.

To compensate for this, the designer needs to make sure that every page in a web site can serve as the front door. It needs to identify which site the page belongs to, be self-explanatory with regards to its functionality, and provide a way to reach any other page in the site.

Novice and visiting users

Users must install Desktop applications and Mobile applications in order to use them. This is not true with Web applications; users only need to click a link to reach a Web application. As a result, the user has a much lower level of commitment with a Web application than with a Desktop or Mobile application. Thus visitors expect to be able to figure out Web applications immediately and the value proposition needs to be readily apparent. If the user is not satisfied in the first few seconds, there likely will not be a second chance.

Types of Pages

To account for the unique challenges of the web and to leverage general usability principles, a set of standards has been developed for Web applications. These standards pertain to the home page, menu pages, and content pages. Each of these pages need to reflect the branding of the site, provide global navigation, and remain focused on content and purpose of the site.

Home page

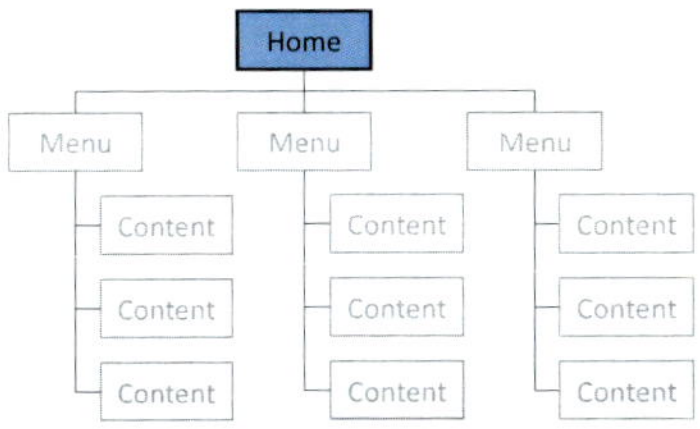

Figure 18.1 - Site map with home page

The home page for a web site has many of the same functions as the lobby to a hotel. Visitor's first impression of the hotel is formed by their experiences in the lobby, leading designers to make the furnishings and decorations of the lobby a cut above the rest of the hotel. Despite this, the lobby is not designed to be a place where the guests dwell. Instead, it is to direct them to the services that they need from the hotel. Finally, a lobby is a communication focal point of the hotel, where announcements are made and where people meet. In all these ways, a home page is like a lobby. It serves the same purposes and needs to have the same attributes. The criteria that all home pages must adhere to are:

Advertise Content A fundamental purpose of a home page is to advertise what the site has to offer. It must be inviting, and it must be a true reflection of the content and functionality of the site. (Mapping, Motivation, Priority 1)

Direct Users A home page must direct users to the content and functionality they seek. It must be easy for them to reach that functionality, ideally in one or two clicks. (Mapping, Efficiency, Visibility, Priority 1)

Interesting A home page should provide something of interest that encourages return visits. (Motivation, Priority 2)

As each home page must represent the organization for which the site was built, each home page by necessity must have a different look and feel. Some home pages are dominated by branding; projecting to the user what the organization is all about. Great examples of this are religious and educational site home pages. Some home pages are dominated by navigation tools providing a means for the user to quickly get to the content or product they are seeking. This is the case for eCommerce sites designed to facilitate product purchasing. Some home pages are dominated by timeliness, providing up-to-the minute updates and dynamic content. News and social networking sites have home pages centered on timeliness to encourage return visits. Finally, some home pages are dominated by the tools the site offers. Service-oriented sites such as search engines and reference sites commonly have home pages dominated by tools.

Menu page

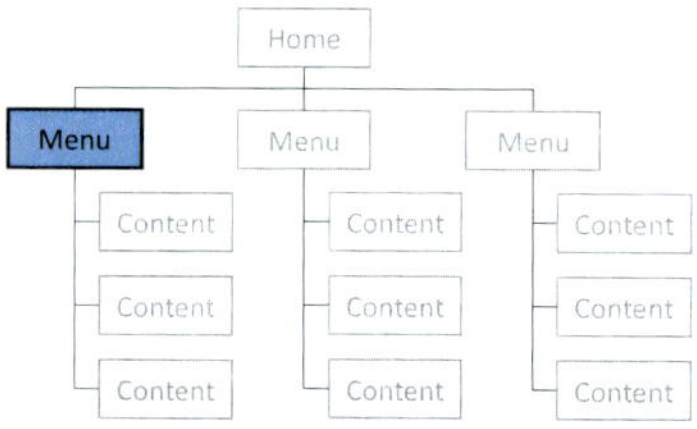

Figure 18.2 – Site map with menu page

The typical web site has anywhere from a dozen pages to several hundred. It is therefore impractical to have a direct link from the home page to every possible content page. The menu page is designed to fill this gap. As the home page is analogous to the lobby of a hotel, a menu page is the foyer or atrium. It represents a sub-set of the larger site, usually tied together by a common theme or purpose. These sub-sets are often called districts. (Lynch & Horton, 2009)

Consider a large city filled with many businesses, homes, and parks. Usually such a city would consist of districts, such as auto row, downtown, the shopping district, and Chinatown. While these districts are still obviously part of the greater whole, they have their own personality. Web sites similarly are sub-divided into districts.

When organizing the content of a web site into a structure intuitive and logical for the visitors, it is often helpful to think about districts. How can this part of the larger site feel cohesive yet also have a personality of its own? What is the common thread to bring all the content and functionality of the district together? The menu page of the site is the home page of the district and is the focal point of the district's identity.

The criteria that all menu pages must adhere to are:

Advertise Content The menu page must inform the user of the types of content contained in the district. Returning users must readily recognize the look and feel of the district and new users must notice that something is different. (Mapping, Motivation, Priority 1)

Direct Users The menu page must direct the visitors to the content page they seek as well as to the main home page (Mapping, Efficiency, Visibility, Priority 1)

Reveal Site Structure Large web sites often have multiple districts and sub-districts. The menu page must project to the user where they are in the context of the larger site and help the user understand the relevant parts of the site architecture. (Mapping, Priority 2)

Observe how there is a large amount of overlap between the criteria for a home page and for a navigation page. Both direct the user to the content below the page in the site hierarchy as well as advertise what exists in those content pages. The difference, however, is that the menu page must also direct users up in the hierarchy to the home page or to higher-level menu pages. Another difference is that, while users may land on a menu page, it is far more common for them to land on and bookmark the home page. Thus menu pages do not have to be as interesting as home pages.

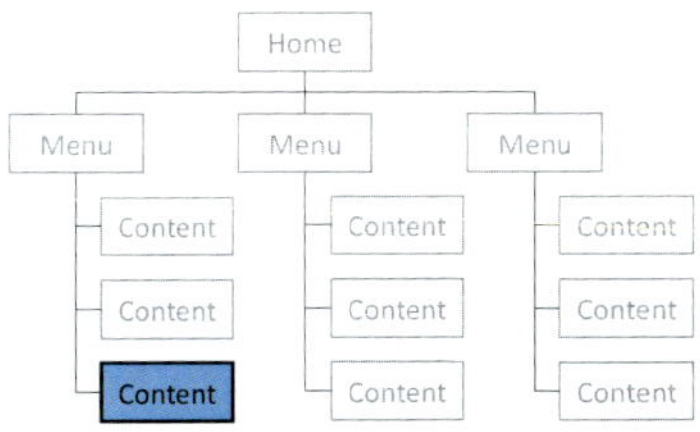

Figure 18.3 – Site map with content page

Content page

The whole point of a web site or a web application is to present the content pages to the user. No matter what factors may influence the page design, it is important to never forget this fact. A content page must remain true to its purpose.

While a web site may have a single home page and a dozen or so menu pages, there are commonly hundreds of content pages. This means that the relationship between the individual content pages becomes important. This brings us back to the layout principles of Unity and Rhythm. Unity states that each content page should look like part of the greater whole. Rhythm states that each content page should be interesting in its own right and not bore the user. How can Rhythm and Unity needs be met simultaneously? The answer can be found in the template design.

Template:

A set of guidelines giving the designer freedom to explore variations in design within a pre-determined set of bounds.

A **template** is a set of guidelines giving the content page designer freedom to explore variations in the layout of a page. The template needs to be rigid enough to convey the branding of the site and meet the Unity needs of the page, yet flexible enough to allow the content page to be true to its purpose as well as provide some variety (Rhythm). Needless to say, template authoring is tricky business best left to a graphic designer.

Template design starts with the content page. This is done because template authoring is most challenging with content pages and also because content pages are the most important pages on the site. Next, the template describes the layout of the menu pages. Each of these should carry through the look and feel of the content page but be obviously different (reflecting their different purpose). Finally, the home page template is built from the menu pages.

The template for a content page, and by extension the content pages built from the template, must meet the following criteria:

Focus on the Content	None of the template features must distract from the purpose of the page. (Visibility, Priority 1)
Honor the Look and Feel of the Site	The content page must look like the rest of the site. The user must be able to tell at a glance what site the content page belongs to. (Visibility, Mapping, Priority 1)
Access to Home	The content page must provide access to the rest of the site, typically with a link to the home page or to the nearest district page. The user must always be able to get to the other pages in the site. (avoid the Gulf of Execution - Intention, Priority 1)

There are several tools at the disposal of the template author. These include the page header, local navigation, primary content, secondary content, and finally the footer.

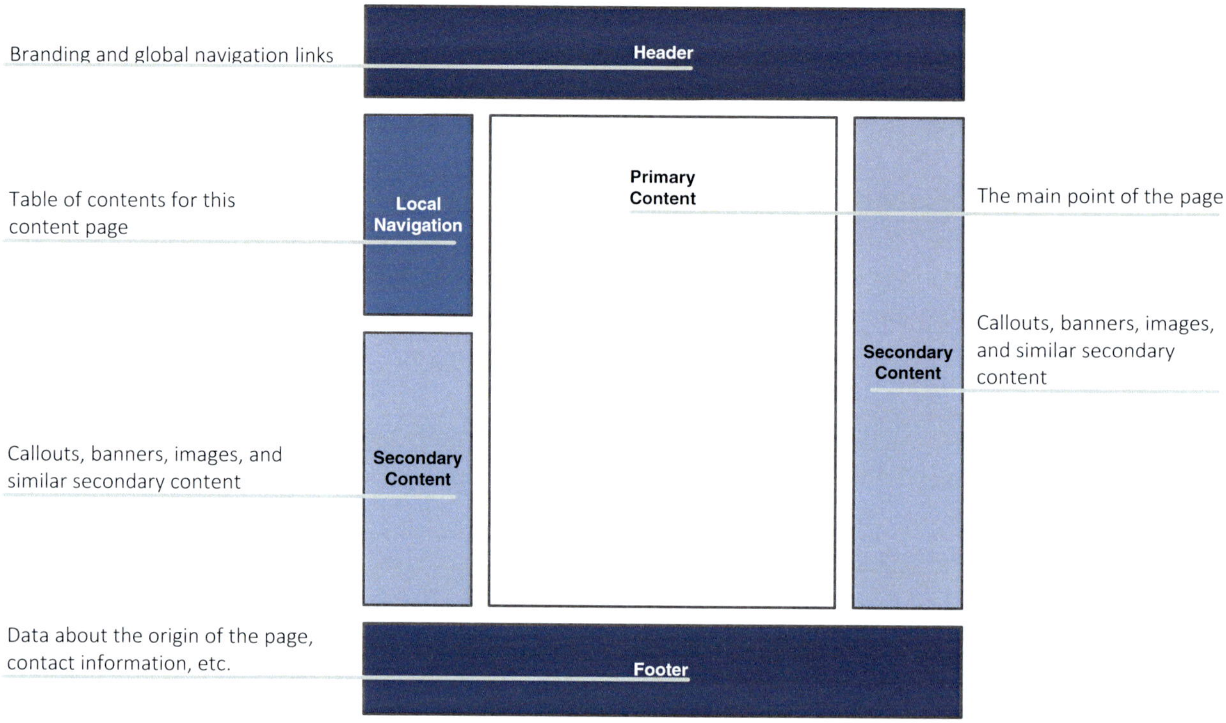

Figure 18.4 - Components of a template

Given a basic layout such as this, it is common to have multiple variations on the same theme, each of which has the same look and feel.

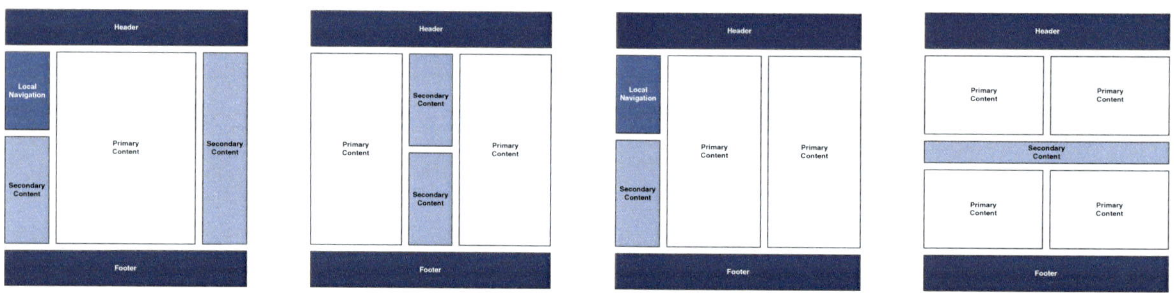

Figure 18.5 – Variations of content template variations

The portion of the template completely independent of the content, the Header and Footer rendered in dark blue above, are also used in the menu page and the home page. Of course, those pages have a much larger header and a much smaller content page. An example of a home page (left) and menu page (right) template are the following:

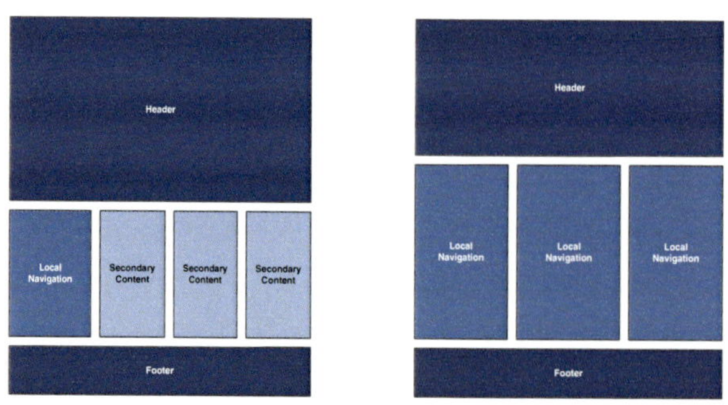

Figure 18.6-7 - Home page and menu page templates

Examples

Example 1 **Home Page**

Analyze the following web page according to the standards for home pages.

Figure 18.8 – Home page for Wyatt's Wanderings blog

Advertise Content The prominent background image and the news feed on the left strongly suggest what this web site is all about: a travel log (blog). Few people visiting this page would be confused as to what is contained herein.

Direct Users The four links in the menu bar are clearly indicating where the user can go for more information. The news feed also appears to be click-able. All these things give the user an obvious place to go.

Interesting The background image changes weekly. The news feed appears to be updated several times a day. Therefore, a returning user is likely to find something new and interesting every time the site is visited.

Example 2 **Menu Page**

Analyze the following web page according to the standards for menu pages.

Figure 18.9 – Menu page for Wyatt's Wanderings blog

Advertise Content This district is clearly about locations. The name of the district (indicated by the bold text on the menu bar) and the background (a topographical map) helps to reinforce this notion. The pictures, however, can be somewhat confusing. Some pictures feature the location prominently. Others seem to emphasize people or activities. A better selection of pictures would help in this regard.

Direct Users The large grid suggests many trips that the user can read about. Each one is represented with a picture. However, the lack of a label makes it difficult to predict exactly what is contained under each of these pictures. As a result, the site could do a better job of directing the user to the specific location he or she would want to learn more about.

Reveal Site Structure The menu bar indicates which district the user is currently in, as well as what other districts exist elsewhere in the site. This helps to reinforce the mental model of how content is organized.

Example 3 **Content Page**

Analyze the following web page according to the standards for content pages.

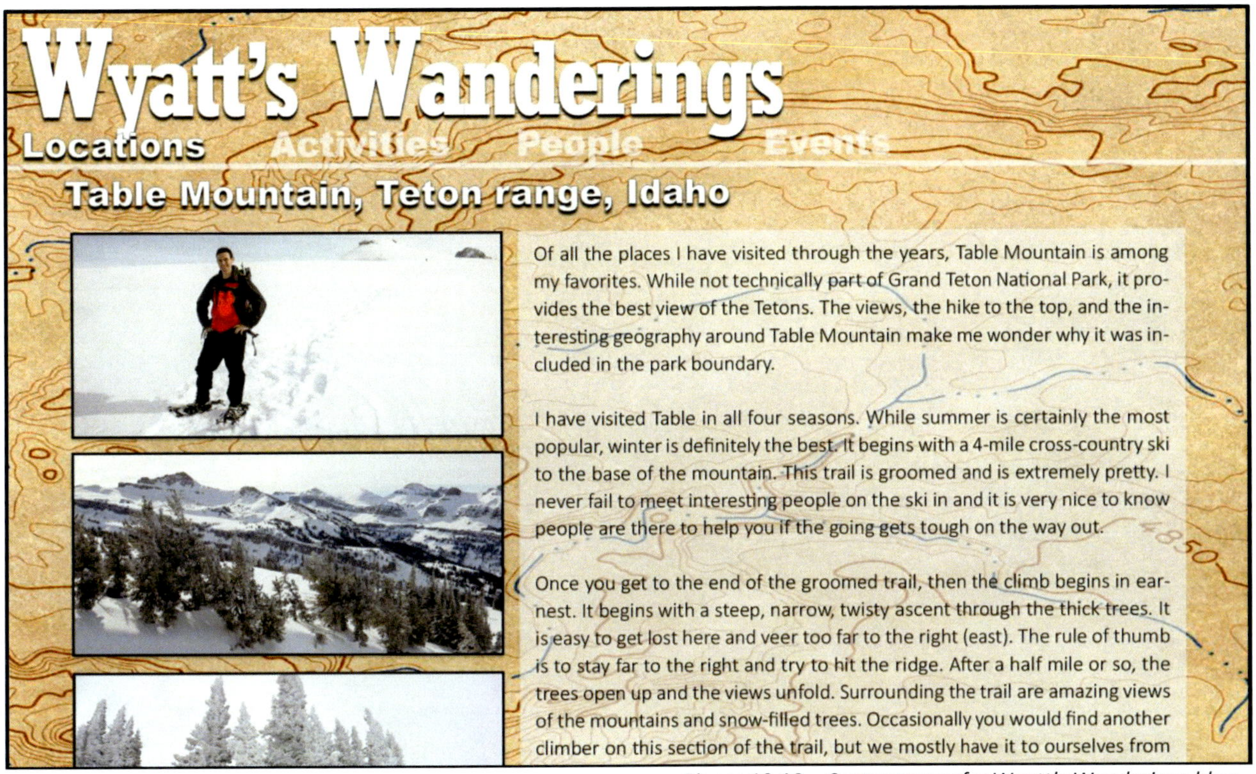

Figure 18.10 – Content page for Wyatt's Wanderings blog

Focus on the Content The vast majority of the real estate of this page is devoted to the pictures and the story associated with this location. However, the connection between the story on the right and the pictures on the left is disjoint at best. It appears that the author was not given the freedom in the template to better mix these design elements.

Honor the Look and Feel of the Site The home page logo, the menu bar, and the district background are prominently displayed in this page. There is no confusion that this page is part of the larger site and part of the district. Such subtle design cues as the picture borders and the velum background on the text are also carried over from the home page and the district pages.

Access to Home The "Wyatt's Wanderings" logo is a hyperlink taking the user to the home page, and the menu bar is clickable taking the user directly to the various districts. The user is therefore confident that this is not a dead end and he/she can always get back.

Comprehension check

Quiz 1 **Districts**

From memory, define a district and describe why one is needed in a large web site

Quiz 2 **Template**

From memory, list the types of things that a web template should specify, and which it should not.

Quiz 3 **Properties of Web Applications**

Which of the following are the unique properties of a web application? Select all that apply

- Large amounts of content are intermixed with the interactive elements
- The application could be used in a wide variety of social contexts
- The user is often highly invested in learning how to use the application
- There is often no indication when the user enters or leaves an application
- The user commonly cannot afford to devote much attention to using the application
- The most complex user tasks are commonly done with this type of application
- The system provides controls that are beyond the control of the programmer
- It is very difficult to input data into the application
- The user has more patience using this type of application than others
- There is a low threshold for using an application requiring little commitment
- There are severe constraints on the output options
- The best visual and sound output is available on this type of application

Problems

Problem 1 **Wedding Photography**
Helfrich Photography is a medium sized studio specializing in weddings, special events, and portraits. The target audience is middle-class to upper-class couples. There is only one use-cases: trying to figure out if "Helfrich Photography" is the right choice for a couple's upcoming wedding. The following is a picture of the home page. The main picture changes every 5 seconds, cycling through a collection of 20 images.

Critique this home page according to the principles of web design. Pay special attention to the criteria for good home page design:

- Advertise Content
- Direct Users
- Interesting

Suggest design alterations that could overcome some of the current design's shortcomings.

Figure 18.11 - Home page to Helfrich Photography sevices

Problem 2 **News Home Page**

Select your favorite news web site. A few choices might be:

- Google News: https://news.google.com/news/
- CNN: https://www.cnn.com/
- NBC News: https://www.nbcnews.com/
- FOX news: http://www.foxnews.com/
- New York Times: https://www.nytimes.com/

Critique this home page according to the principles of web design. Pay special attention to the criteria for good home page design:

- Advertise content
- Direct users
- Interesting

Suggest design alterations that could overcome some of the current design's shortcomings.

Problem 3 **News Menu Page**

Start with the news site from Problem 2. Navigate to one of the section pages such as Sports, Politics, or World. Critique this page according to the principles of web design. Pay special attention to the criteria for good menu page design:

- Advertise content
- Direct users
- Reveal site structure

Suggest design alterations that could overcome some of the current design's shortcomings.

Problem 4 **News Content Page**

Start with the news site from Problem 2 and 3. Navigate to one of the articles. Critique this page according to the principles of web design. Pay special attention to the criteria for good content page design:

- Focus on the content
- Honor the look and feel of the site
- Access the home page

Suggest design alterations that could overcome some of the current design's shortcomings.

19 Mobile Applications

Mobile applications are defined as programs running on a device meant to be operated while standing. A tablet is therefore a mobile application because it is easy to hold the device with one hand and operate the device with the other. If a keyboard is attached to the device, it becomes functionally very similar to a laptop. However, because the apps still behave in the same way with or without the keyboard, it remains essentially a mobile device.

While mobile devices began in the 1980s with dedicated PDAs (Personal Data Assistants), they did not really take off until integration with mobile phones became more seamless. Today mobile apps have become equally important as web apps and traditional desktop apps in terms of both market validation and value to end-users.

Attributes of mobile computing

There are several things that set mobile applications apart from their Web and Desktop counterparts. These are the social context, user patience, input constraints, and output constraints.

Social context

Mobile applications are meant to be used on-the-go. This means that you are likely to check your cell phone or tablet while standing beside someone. Mobile devices find their way into conversations and are easy to pass to other people.

Developers of mobile applications need to realize that their applications do not have all of the user's attention. They must not assume this attention and must not strive to get the user's attention; instead they should contribute to the user's conversation and his social context.

User patience

Users of a Desktop or Web application are typically willing to spend a couple minutes to get the information they need. This is because, in both cases, they are sitting and the application has the user's undivided attention. In many mobile scenarios, this is not the case. The user turns to his mobile device to get a piece of information necessary for their social context. This typically means the information must be received immediately.

Mobile users are less patient with their software because their primary task at that moment is not "interacting with their computer." Developers of mobile applications need to realize this and design their apps appropriately. Perhaps the United States Marine Corps saying of "Be blunt, be brief, and be gone!" applies here.

Input constraints

Mobile devices are not optimized for efficiency of data entry like desktop computers are, instead optimized to be, well, mobile. Thus input-centric activities such as word processing are less important than output-centric activities such as reading news.

While modern devices have a rich vocabulary of input options available, they will always be inferior to desktop hardware simply because desktop hardware has fewer constraints put on their design. To compensate for this, great care needs to be taken when designing input interfaces.

Output constraints

As with input devices, output devices for mobile devices are often severely limited by size, weight, and battery constraints of the device itself. It is very difficult to produce rich sound from a mobile device without headphones. It is very difficult to provide a large screen without severely influencing the portability of the device. It is very difficult to provide the processing power to do rich graphics without severely limiting battery performance.

Mobile developers need to adopt an entirely different perspective towards display than Web or Desktop developers. Screens need to contain less controls to fit smaller footprints, layouts need to be simplified so the user can understand them in a glance, and the developer cannot rely on sound to communicate important information.

Standards

The most important variables of usability for a mobile app are typically efficiency (mobile users have severe input constraints and need to gather information quickly from their device), visibility (mobile displays are commonly cramped and the user must get information at a glance), familiarity (the app should be consistent with other apps on the platform), and motivation (do not forget the social context). Taking these considerations into account the following general standards and guidelines apply to most mobile design problems: honor the platform, minimize user effort, be succinct, use finger-sized inputs, and take performance constraints into account.

Honor the platform

Every application on a given mobile platform should honor the standards of that platform. In other words, it is generally a bad idea to port an app from one platform to another without taking into account the paradigms, look & feel, and standard controls of that platform. The designer must carefully read and internalize the standards to that platform:

- **Google Android:** http://developer.android.com/design/index.html
- **Apple iOS**: http://developer.apple.com/library/ios/documentation/UserExperience

Minimize user effort

In order to make sure that the app can be used in the mobile context, it must be handled with the minimal amount of user effort. This means that input-heavy scenarios are unlikely to work well with a mobile application. When input is absolutely required, great pains must be taken to simplify the task as much as possible. In order to accomplish this:

- **Balance input requirements with value**. The longer the application asks the user to toil over a touch keyboard, the more the application should offer the user in return. If there is no immediately apparent user benefit to his input effort (such as scrolling through an endless End User License Agreement (EULA) or closing popup adds in a video sharing service), the user is likely to become quickly discouraged and quit the app.
- **Simplify choice selection**. There are often many ways to facilitate the user making a selection. For example, if the application needs the user to choose a month of the year, the app could have the user type the month name, scroll through a list of months, or tap on a month name from a 3x4 grid of months. Of these, clearly the grid option is the easiest for the user. With mobile applications, a considerable amount of effort on the part of the designer should be spent simplifying such selections for the user.

Be succinct

In the mobile scenario, efficiency and visibility considerations are paramount. The user can only glance at her device for a moment to gather needed information or make a decision. Therefore, the application must be able to both give answers to the user immediately and reward more through scans with details. There are several methods at the designer's disposal:

Worst Use sound. Sound is often used to convey information on a desktop application because it does not take any real-estate and there is almost always a speaker connected to the machine. In the mobile context, however, sound should almost never be used: it can be unexpected and embarrass the user.

Bad Use dense text. Lots of text requires large amounts of attention on the part of the user. Not only can this be annoying and less efficient, it could be dangerous. The user could be walking across the street while using the app.

Acceptable Use a short label. A short label of one or two words can be quickly scanned and provide the user with the most important information. This is especially true if the information is displayed prominently and is easy to separate from the rest of the design.

Best Image. Images are easy to locate on a design and, if done correctly, are easy to understand. A familiar symbol or descriptive image is often the best way to convey information to the user.

Finger-size inputs

With a desktop application, mouse hit-regions can be as small as a few pixels. Most users have remarkable dexterity using a mouse. With mobile devices, however, the human finger is a blunt instrument. If the required hit region is smaller than a centimeter, then an Execute Sequence gulf of execution is likely to result. So, how big must the hit region be? There are three variables in this equation: the number of pixels per inch, the size of the finger, and the accuracy of the display finding the center of the finger (Anthony T, 2012).

- **Pixel density**. This is commonly measured in pixels per inch (PPI) (analogous to dots per inch (DPI) on a printer), the number of pixels per linear inch in a display. The first generation mobile displays were 100ppi, the first iPhone was 163ppi, and the first iPad was 132ppi. As a reference point, most laptops and LCD desktop displays are 120ppi. When Apple introduced the "retina display," a display with a pixel density a little sharper than the human eye can resolve in the best circumstances, 326ppi was chosen. The designer of a mobile app must know the pixel density value for a device in order to be able to accommodate finger-sized inputs.
- **Size of the finger**. The average human finger size is 1.6cm to 2.0cm for adults (Dandekar, Raju, & Srinivasan, 2003). That being said, many users incorporate the thumb (especially for games) which is 2.5cm on average. In other words, the designer must take into account the finger used (based on expected usage and position on the screen) and the type of user (child or adult).
- **Accuracy of the display**. Most mobile operating systems have an API allowing the developer to know the center of a tap within a few pixels. In other words, the user's finger on the display yields an oval-shaped image on the device's sensor. The shape of the image depends on how hard the user presses on the display and how accurate the sensing technology. Most mobile operating systems (iOS, Android, and Windows Phone) have a simple algorithm that finds the center of the finger image and registers it as the tap location.

How do these three variables interact? Consider a simple button on Apple's iOS platform. The pixel density is 326ppi on the iPhone5. The average finger size is 1.8cm, translating to 0.709" or 231 pixels. This corresponds to how many pixels the typical adult finger covers on the iPhone5. According to the iOS User Experience Guidelines, a tap-able element should be 44 x 44 points (a point is a device-independent measurement, corresponding to a single pixel on a 163ppi display) or 88 pixels. Therefore, Apple claims that the accuracy of their display is 1/3 finger size (38.07% to be exact). The equation is:

$$hitRegionSize = fingerSize * accuracy$$

Another example is the Nokia Lumia 1020. The device has 334ppi and an accuracy of 1/4 finger size. My finger is 1.7cm. The hit region must be:

$$\begin{aligned} hitRegionSize &= fingerSize * accuracy \\ &= 1.7cm * 0.25 \\ &= 0.6692'' * 0.25 \\ &= .1673'' \\ &= \lceil 55.878\ pixels \rceil \\ &= 56\ pixels \end{aligned}$$

Performance is important

In the mobile context, users are almost always doing something else while interacting with an app. They could be in a meeting, walking, waiting in line, talking with friends, or a thousand other things. In each case, they expect any time spent interacting with their mobile device to be offset with some benefit. This is definitely not the case when the app is taking a long time to load or respond to a request.

Performance must be nearly immediate on a mobile device. A long delay could be embarrassing for the user if she is trying to share something with a friend. An unexpected lag could be dangerous if the user is engaged in a time-sensitive activity (such as walking) in the physical world. Large amounts of processing or network interaction can also starve other apps running on the device and drain the battery.

In most cases, it is better to have less functionality on a mobile app than suffer from poor performance. Less functionality might disappoint the user, but poor performance may make the user angry. This often requires the designer to drastically simplify the design.

Examples

Example 1

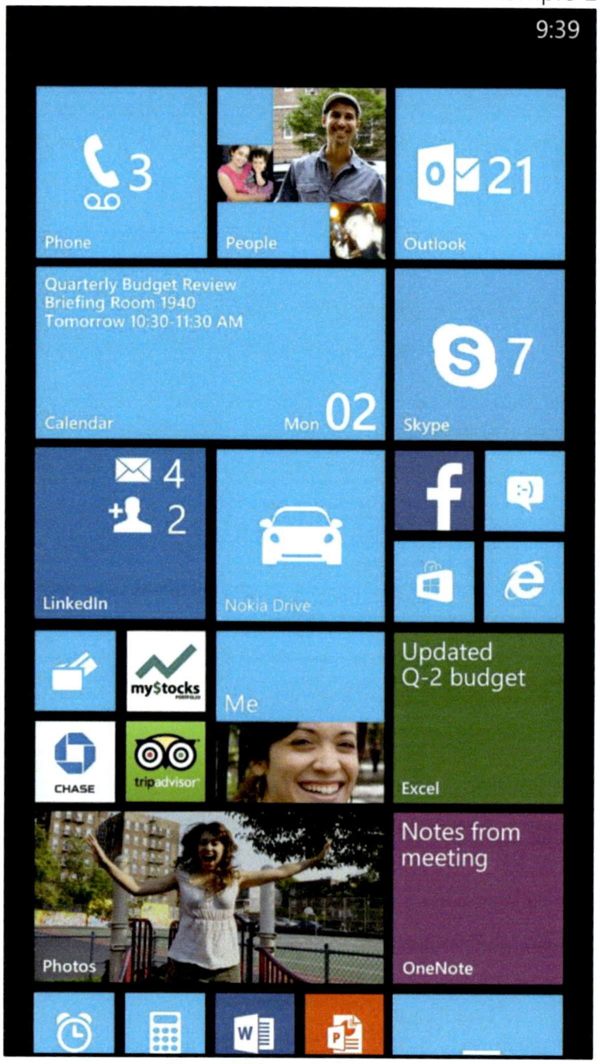

Figure 19.1 - Windows Mobile 8 home screen

Windows Mobile 8

Microsoft developed the Windows Mobile operating system to compete with Apple's iOS and Google's Android. Critique the interface design according to the 5 components of mobile user interface: honor the platform, minimize user effort, be succinct, finger-size inputs, and performance.

Platform

Microsoft has 9 platform standards. One is "Layout: Get on the grid" or to use a grid for layout. Here we can see that there is a 6-column grid with rows of exactly the same size. Thus the grid standard is honored. Another is "Infographics: Info is in." Here all of the tiles use infographics except for some with pictures. Though you could argue that the pictures should be removed, another standard is "Let your content breathe" which is what this design does. Overall, this screen honors all of the standards.

Effort

Only one tap is necessary to get 27 applications and a single swipe doubles the number. Furthermore, a great deal of information can be reached without even touching the interface (such as the next item on the calendar).

Succinct

The same arguments as "Effort;" much of the effort wins occur because the interface is so terse.

Finger-Sized Inputs

The resolution of this device is 312dpi, the finger sensitivity is ½ finger, and the finger size is 1.7cm. This yields a minimum hit region of 105 pixels. Each tile is 248 pixels so the standard is reached.

Performance

The tiles update whenever the phone is on which does consume battery performance. However, the tiles themselves are easily drawn so they mitigate the load on the battery.

Example 2

Figure 19.2 - Scrapbook mobile application

Scrapbook Application

Consider a mobile application designed to share images and stories. The user takes a picture from his or her phone and then adds a few words describing what happened. The application then creates a simple scrapbook-style page to be viewed or shared. This application is for Apple's iOS 7 platform.

Platform

Apple's iOS Best Practices has several guidelines which are relevant here. The first is "focus on the primary task." This application meets the guideline because the vast majority of the screen real-estate is devoted towards the user's content. A guideline for iOS 6 was "Consider adding physicality and realism." The photo-corners and textures, and drop-shadows accomplish this goal. Note that iOS 7 removed this guideline so the design no longer honors the platform.

Effort

All the controls can be reached in a single tap.

Succinct

All the actions are represented with a single icon, making the interface succinct to the point of being too terse.

Finger-Sized Inputs

This application targeted the iPhone 5 with a resolution of 1,136 x 640 pixels and a density of 326 pixels per inch. With an average finger size of 1.7cm and accuracy of 1/3 finger size, the hit region should be (1.7cm * 0.33 = 0.6692" * 0.33 = 0.2201" = 72 pixels). Each icon hit region is 88 x 88 pixels so the standard is met with 16 pixels to spare.

Performance

The page is mostly static unless responding to user action so the application has little impact on battery or wireless resources.

Comprehension check

Quiz 1 **Mobile Guidelines**

For each of the following, identify the corresponding mobile application design guideline:

- People have more dexterity with a mouse than they do with their finger
- All the apps on a device should feel like they belong together, parts of a greater whole
- It is difficult to type while walking
- The user needs to be able to get the information she needs while walking down the street talking to a friend
- Mobile devices have small batteries, small amounts of memory, and limited processors

Quiz 2 **Properties of Mobile Applications**

Which of the following are the unique properties of a mobile application? Select all that apply

- Large amounts of content are intermixed with the interactive elements
- The application could be used in a wide variety of social contexts
- The user is often highly invested in learning how to use the application
- There is often no indication when the user enters or leaves an application
- The user commonly cannot afford to devote much attention to using the application
- The most complex user tasks are commonly done with this type of application
- The system provides controls that are beyond the control of the programmer
- It is very difficult to input data into the application
- The user has more patience using this type of application than others
- There is a low threshold for using an application requiring little commitment
- There are severe constraints on the output options
- The best visual and sound output is available on this type of application

Quiz 3 **Template**

I am writing a children's app for the original Kindle Fire. The device is 169ppi, the size of a child's finger is 1.6cm, and the accuracy of the display is ½ a finger. What is the minimal control size in pixels for the Kindle Fire. Just write the number rounded up such as "56"..

Problem 1

Figure 19.4 - Travel application for Andriod

Travel agent application for Android

A travel agent has produced an Android mobile application to share a trip itinerary with the clients. This application presents a brief summary of each day with a representative picture. Critique this application according to the standards for mobile applications. Pay special attention to:

- Honor the platform
- Minimize user effort
- Be succinct
- Finger-size inputs

Suggest design alterations that could overcome some of the current design's shortcomings.

Problem 2

Calculator application

A calculator application has been created for the Android system which facilitates computing several common tasks. Critique this application according to the standards for mobile applications. Pay special attention to:

- Honor the platform
- Minimize user effort
- Be succinct
- Finger-size inputs

Suggest design alterations that could overcome some of the current design's shortcomings.

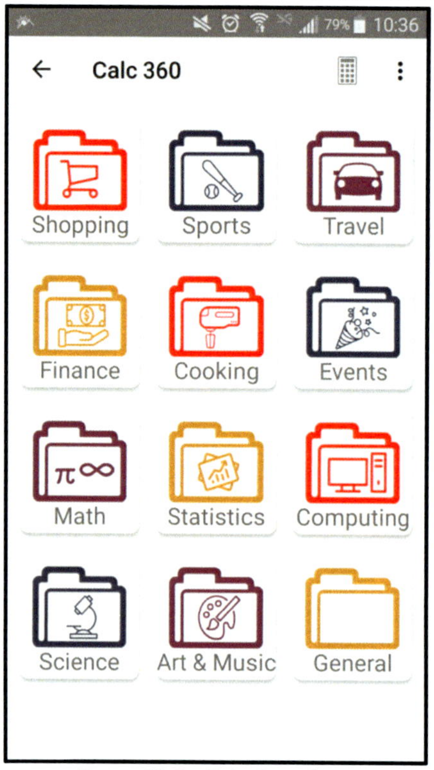

Figure 19.6 – Calculator home screen

Figure 19.7 - Calculator finance screen

Problem 3

Week Calendar for iOS

Critique iOS application "Week Calendar" app according to the standards for mobile applications. Pay special attention to:

- Honor the platform
- Minimize user effort
- Be succinct
- Finger-size inputs

This version of the application is built for the iPhone 8.

Suggest design alterations that could overcome some of the current design's shortcomings.

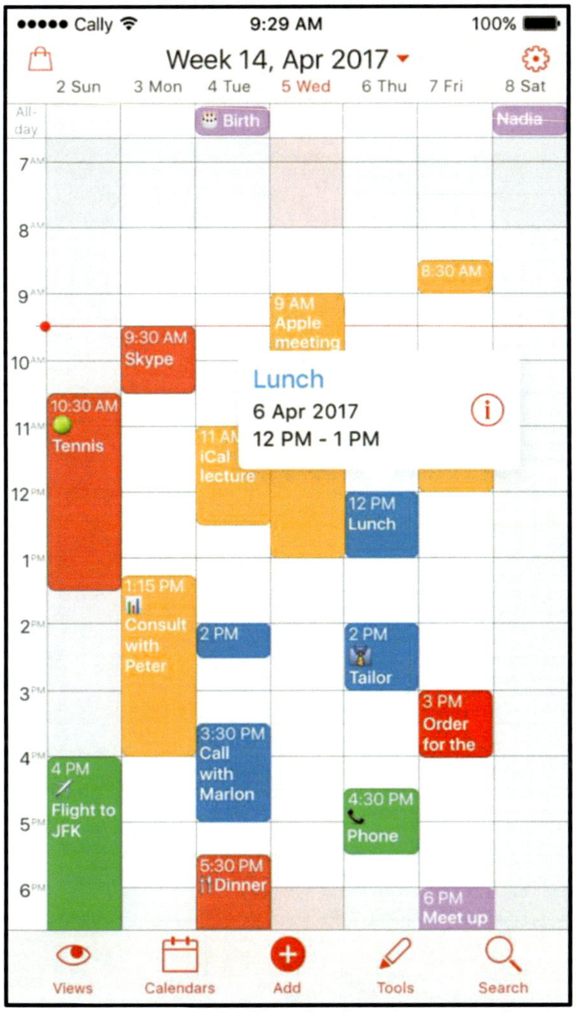

Figure 19.3 - Weekly Calendar iPhone 8 application

Problem 4 **Excel for iOS**

Critique Microsoft Excel for iOS according to the standards for mobile applications. Pay special attention to:

- Honor the platform
- Minimize user effort
- Be succinct
- Finger-size inputs
- Performance is important

Suggest design alterations that could overcome some of the current design's shortcomings.

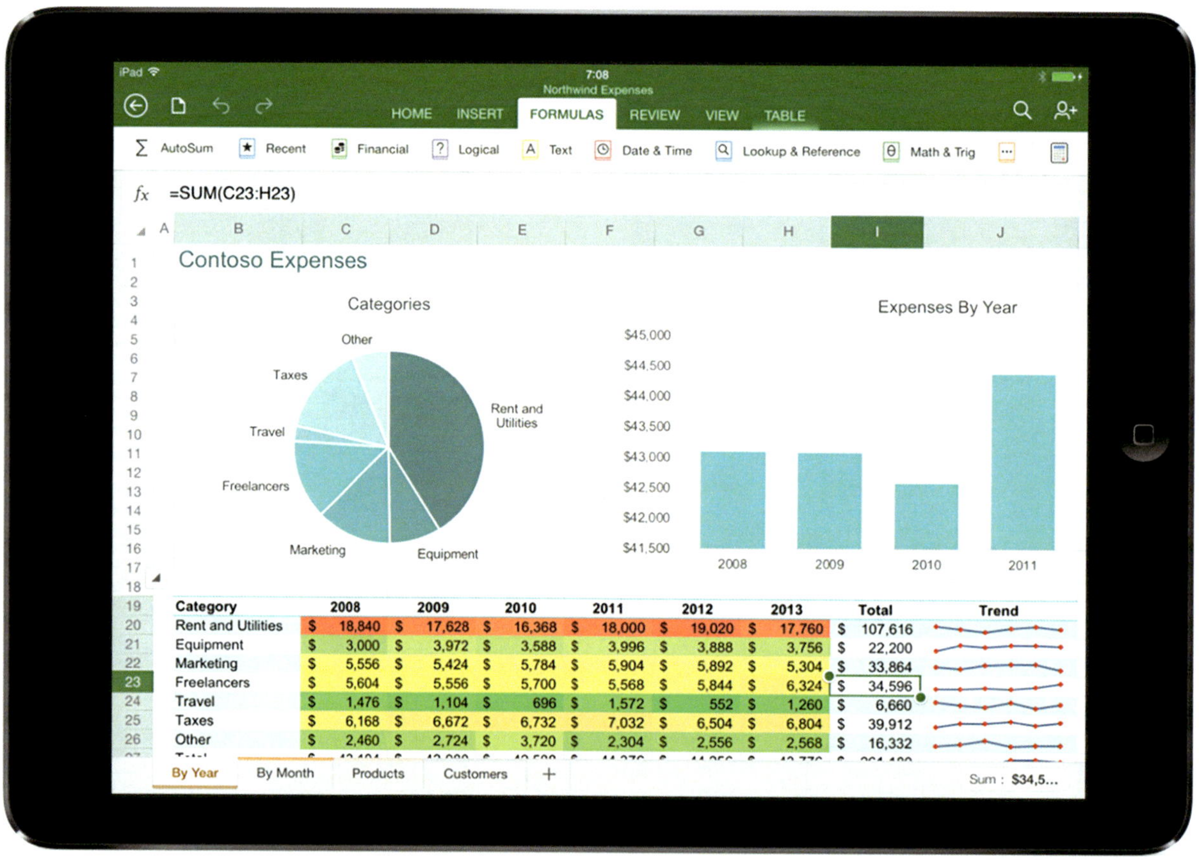

Figure 19.4 – Microsoft Excel for Apple's iPad platform

Problem 5 **Car Dashboard**

Critique the dashboard to a car that has an "all glass" display. All of the controls are reachable through a large touch screen mounted between the driver's and passenger's seat. Pay special attention to the scenario as discussed earlier in the book. Analyze his design according to:

- Honor the platform
- Minimize user effort
- Be succinct
- Finger-size inputs

Suggest design alterations that could overcome some of the current design's shortcomings.

Figure 19.9 – Dashboard to a car with an "all-glass" cockpit

20 Desktop Applications

Desktop applications are defined as programs running on a computer designed for seated use that appears to the user to be a stand-alone application. There are two components of this definition. First, desktop applications are designed to be used by individuals giving their undivided attention to the task at hand. The user is typically seated in front of a screen with rich input and output options. Second, each desktop application must appear as a separate entity on the user's operating system. Therefore the defining characteristic of desktop applications is not the hardware on which the program runs (laptop computer, desktop computer, or even tablet computers), not the technology on which the application is built (it could be developed in HTML and presented in a borderless window or it could be installed on a tablet), nor the task the user is trying to accomplish. Instead, it is a function of how the user perceives the application: a stand-alone tool requiring undivided attention.

Initially all applications were desktop because desktop computers predate mobile and web technology. Desktop applications were written on million-dollar computers and operated by skilled technicians. As personal computers began to permeate the home markets, personal productivity and entertainment applications were developed to meet the needs of this new user base. Similarly, the skill level required of the users steadily dropped. No longer were trained professions needed to operate a desktop application, now a child can use one. Still the important characteristics of desktop applications remain: a user sitting in front of a computer focusing on one task at a time.

Attributes of desktop computing

There are several attributes that set desktop applications apart from their Web and Mobile counterparts. These are the investment the user is willing to place in these applications, the degree of complexity of user tasks, the degree of user patience in performing these tasks, and the richness of input and output communication channels.

Investment

The typical user may visit hundreds of web sites in a month, have dozens of apps installed on their mobile device, and yet only use only a half a dozen desktop apps. This same user will balk at spending any money for the privilege of using a web site, grudgingly spend a few dollars on a mobile app, and be willing to spend several hundred dollars on a desktop app. Users are often loyal to their desktop apps, using the same app for a dozen years through several versions. It is not uncommon for a user to buy a book or take a class on how to use Adobe Photoshop or Microsoft Excel. They are willing to do this because they believe their investment will pay off in increases in proficiency or efficiency.

User tasks

Most users reserve complex tasks for desktop applications. While it may be technically possible to write a report on a mobile device, few users would choose a phone over a laptop for writing. These complex tasks often require complex software. Desktop apps are commonly an order of magnitude more powerful (and complex) than their Web counterparts which, in turn, are an order of magnitude more powerful than their Mobile counterparts.

User patience

Users are typically more patient with their desktop apps than they are with their mobile apps. This patience can be traced to several sources:

- **Distractions**: There are fewer distractions when sitting at a desk working on a desktop computer than while walking around with a mobile device
- **Complexity**: Users appreciate that the task they are performing is more complex so they are willing to spend more time doing it
- **Familiarity**: Because users have more invested in their desktop apps than the mobile or web counterparts, they do not mind as much spending a few minutes figuring things out

Communication channels

Desktop applications, as a rule, have the highest quality input and output devices. The screens tend to be the largest and brightest available. Native desktop applications are able to communicate with the operating system and graphics engine to provide the richest displays. Data entry can be high fidelity and seamless. There is no better showcase for this than comparing web and mobile games against desktop and console games.

Standards

Desktop applications are typically built from buttons, menus, edit controls, modal dialogs, modeless dialogs, and the document surface.

Buttons

A button is a clickable region that launches some action. This action frequently changes the application context, either navigating to a different location in the product or returning from a location in the product. In almost all cases, buttons are actions so button names should be verbs. There are many variants of buttons:

Options...

Labeled Button A button set apart with a rectangle or circle. They are designed to look like the buttons found in the physical world, often with visuals making them look three dimensional. Variations of labeled buttons include command buttons (standard button), lightweight command buttons (standard button without a button frame), menu button (drops a menu), and split button (half command button and menu button).

Hyperlink

Hyperlink Clickable text, usually with an underline or a color. This was introduced with HTML by Conseil Europeén pour la Recherche Nucléaire (CERN) in 1989. Hyperlinks are laid out like standard text and blend into text better than other buttons. They are used mostly for navigation.

Toolbar Button Same as a button except usually drawn with an icon instead of with text. They are frequently used to show state, reflecting a single Boolean value (such as the Bold button) or more complex data (such as the font color button).

☑ checkbox

Checkbox A small button with associated text used exclusively to reflect state. This means that checkboxes are not verbs like other buttons, instead their labels are nouns. Checkboxes have more real-estate for label text than other buttons.

Hit Region A part of an image that can be clicked. The image needs to provide a clue that the hit region exists and what action will be performed.

There are many guidelines associated with buttons depending on the system in which they are used and the functionality the button needs to represent. That being said, there are a few guidelines that hold true in a wide variety of situations.

- **Take time to get the label right**. The designer needs to communicate to the user what the button does in just a few well-chosen words. If the user is surprised by the action initiated by a button click, then the label needs to be reworked. The label also needs to be very terse. There is not much real-estate available to describe the functionality behind the button. In most cases, button labels are two words: [Verb Noun]. The verb describes what will transpire and the noun will describe the entity to be affected. Often the noun can be omitted because it is implied by the context. For example, the [Close] button does not need to be called [Close Dialog] because it is implied that the dialog will be the recipient of the action.

- **Use the correct variant**. Buttons that bring up dialogs have a label with an ellipse [Open...]. Buttons that drop a menu have an arrow [Select ▼]. Buttons that expand a window or dialog have chevrons [More >>]. Checkboxes reflect state and have noun labels. When designing an interface on a given platform, make sure you are well versed with all the available button variants and the guidelines for their use. Please see the Apple OSX Buttons guidelines and the Microsoft Desktop Command Buttons guidelines for a reference.

Menus

A menu is a list of related buttons. Frequently a menu is scrollable and often they show state. Types of menus include:

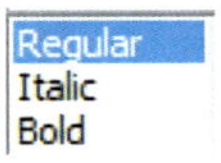

Listbox A set of options usually represented with noun text labels. Listboxes typically show state and are designed to facilitate the user selecting a single interval value.

Multi-select Listbox Same as a listbox except multiple items can be selected. Selection state can be indicated by shading or highlighting the rectangle containing items or with an icon such as a push-pin.

Radio Group A list of checkboxes where exactly one value can be selected at a time. A radio group is functionally equivalent to a listbox but provides more room for descriptive text and is not bounded with a listbox control rectangle.

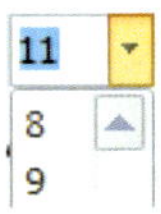

Combo-atomic A listbox coupled with an edit control. Combo-atomic Listboxes give the user the option of making a selection by clicking on a list or typing the name of the entry. This is particularly useful when there are a very large number of possible items, such as a font selection control.

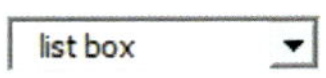

Dropdown Listbox Same as a listbox, multi-select listbox, or combo-atomic listbox except the contents of the list are not presented to the user until the user clicks a "reveal" button [▼].

As with buttons, menus have a wide variety of standards and recommendations depending on the use of the control and the context in which it is used. General guidelines applicable to all menus are:

- **Give the user an appropriate number of selection options**. The human brain is most efficient at making a selection when there are between 5 and 9 options. This is related to the human capacity to keep 7 +/- 2 items in short-term memory at a time (Miller G. , 1955). Therefore, people are much more efficient and accurate selecting from a list of 64 items when an 8x8 menu is presented (8 items, each of which consists of a menu of 8 items) than a 64x1 or 4x4x4x4 or 2x2x2x2x2x2x2x2 menu structure (Miller D. , 1981).
- **Take time to get the label right**. Typically there is more real-estate with menu labels than with button labels. This means that more than one or two words are commonly possible. That being said, make sure the user will understand the chosen words, framed in the context of the user's scenario and avoiding terminology with which the user may be unfamiliar.
- **Provide defaults whenever possible**. Based on what is known about the user and the critical use-cases, make an educated guess what option the user will probably select.
- **Use the correct variant**. As with buttons, this requires the designer to be familiar with all the various flavors of menus available on the target system. Please see the Apple OSX Selection Controls guidelines and the Microsoft Desktop List Boxes guidelines for a reference.

Edit Controls

An edit control is a control allowing the user to provide free-form or loosely-constrained input through the keyboard. While the most common form of edit control accepts plain text, many other variants exist:

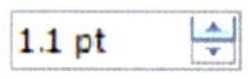

Numeric Input Also known as numeric text boxes or spinners, numeric input controls are designed to facilitate selection of scalar values. If there are not many valid values, use a listbox or another menu control.

Text Box Also known as a text input or simply as an edit control, a text box is a control allowing the user to type in a single line of text. Variants include a token field (grouping user input into enclosed tokens such as file names or e-mail addresses), password fields (all user input is rendered as a dot or an asterisk), and search field (with an MRU auto-complete).

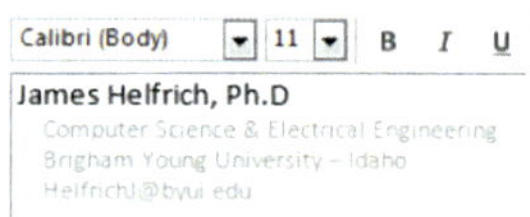

Figure 20.1-3 - Examples of edit controls

Rich Text A control allowing the user to both input text and the formatting associated with the text. These behave much like a document surface for a word processor except with much less real-estate and editing controls allowed.

By far, the biggest and most common mistake that can be made with an edit control is to use one where another control would be more appropriate. Users generally hate edit controls. They often ask the user to provide information that they do not have in a format they do not understand. If an edit control can be replaced with a menu or a button, then it almost always should. Guidelines surrounding edit controls are:

- **Make the question clear**. Every edit control is asking a question where the user is to provide the answer. Make sure that question is clearly stated.

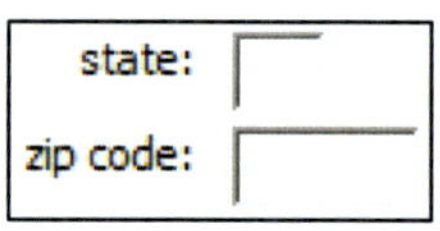

Figure 20.4 - Edit control with clear expectations

- **Make the expectations clear**. Often a specific format of input is required. This is certainly the case with a filename or the name of a typeface. When such a format is needed, make sure that it is explicitly stated so the user knows what is needed. Additionally, when a large amount of text is expected, provide a large amount of space in the edit control. On the other hand, if the edit control is for a state abbreviation, provide only room for two characters of input.

Figure 20.5 - Edit control facilitating user input

- **Provide help**. Consider an edit control allowing the user to input a date. Of course a specific format will be needed; not just any string can be parsed into a valid date. In cases like this, it is helpful to provide a secondary control to facilitate the selection. For example, a date control can be attached to the date edit control allowing the user to either type the date or use a menu. Another example is a file control, allowing the user to either type the filename or select it from the system file browser.

Edit controls can be a very efficient way to enter input because users are often more efficient at using the keyboard than using the mouse. For example, never do the following:

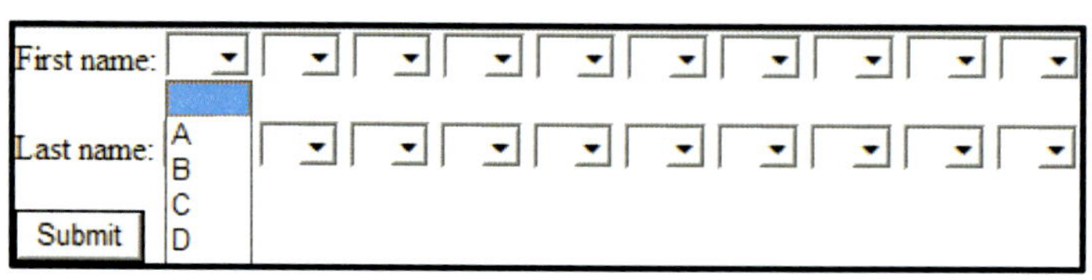

Figure 20.6 - Menu provided when an edit control would be a better choice

Modal Dialogs

A dialog is a window associated with an application designed to help the user perform some task. There are two flavors of dialogs: modal and modeless. A modal dialog is a dialog that sits on top of the application frame rendering the rest of the application inert until the dialog is dismissed. This is appropriate for circumstances when the application needs a piece of information from the user before any action can be taken. Modeless dialogs, on the other hand, do not take focus.

Modal dialogs are one of the most commonly used and abused interface constructs in the desktop application environment. The reason for both is the same: demands are made to the user on the system's terms. This is a very system-centric not user-centric perspective, one easy to program but not very user friendly. As a general rule, users hate dialogs. Dialogs demand information expressed in a terse and technical language with often frightening and unseen consequences. To mitigate these problems, the following guidelines have been developed:

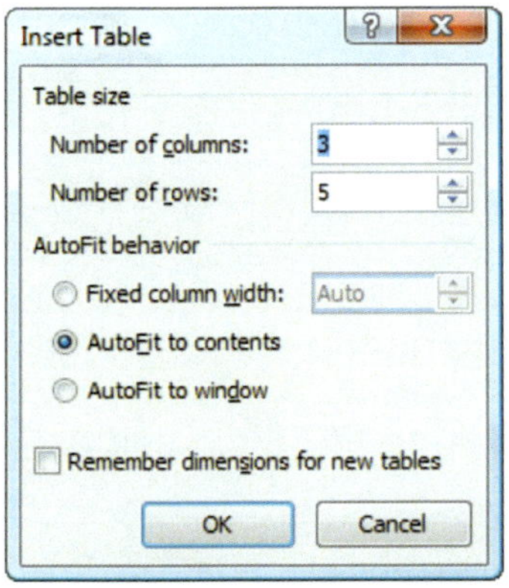

Figure 20.7 - Modal dialog

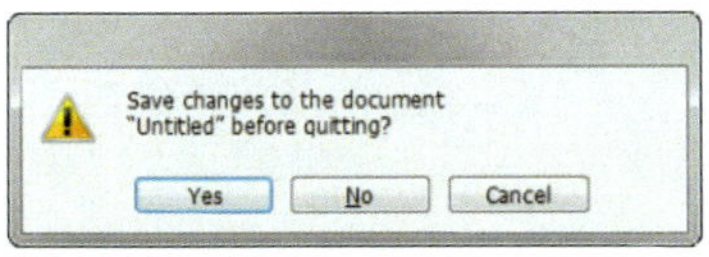

Figure 20.8 - Accept-Reject

- **Be intentional about grouping**. Related and similar controls should be in close proximity to each other, dissimilar controls should not. Dialogs offer a two-dimensional canvas on which to communicate to the user the state of his data and accept input in which his intention is expressed. Utilize this canvas to its fullest extent.
- **Use modal dialogs sparingly**. Only use a modal dialog when there is no other recourse. Examples include critical events that must be dealt with immediately (such as a failed auto-save requiring a new location to be specified to preserve the user's data), a user-initiated action requiring non-trivial interactions (such as printing a document), and peripheral tasks orthogonal to the main purpose of the application (such as adjusting preferences or configuration settings).
- **Make "accept" and "reject" consequences clear**. Users are often confused what they are agreeing to when they click on the [OK] button. Similarly, they often cannot tell the difference between [Cancel], [Close], and hitting the [x] control on the dialog frame. The designer needs to make the purpose of the dialog so clear that it is very difficult for the user to misconstrue the consequences of his actions. In the following example, it would be better if the options were [Save] [Don't Save] [Return to Photoshop]
- **Be intentional about reading order**. Enumerate a few common use-cases for the dialog. For a given design, draw lines connecting the various controls the user must interact with to complete the use-case. A well-designed dialog has a linear reading order, suggesting an efficient path for the user's eyes and cursor to follow to complete his task.

Figure 20.9 - Example of controls arranged in a human-readable way

Modeless Dialogs

Modeless dialogs are similar to modal dialogs except they do not demand the user's immediate attention; the user can continue working while a modeless dialog is up and deal with it at his convenience. There are several variations of modeless dialogs:

Traditional Modeless Dialog

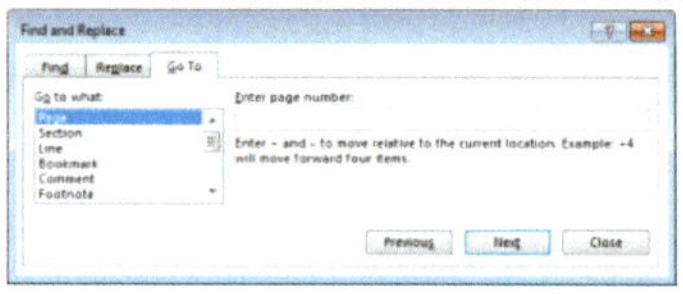

Figure 20.10 - Modeless dialog

A modeless dialog framed in operating system dialog chrome similar to that of a modal dialog. This chrome provides the enclosing grouping mechanism, a title, and common controls (close, help, etc.). Traditional modeless dialogs cannot be docked and float freely over the window from which they were spawned. Sometimes they are tied to the main application frame, disappearing when the application is no longer in the foreground. Sometimes they are tied to the desktop window, accepting focus independently of the application spawning it.

Palette

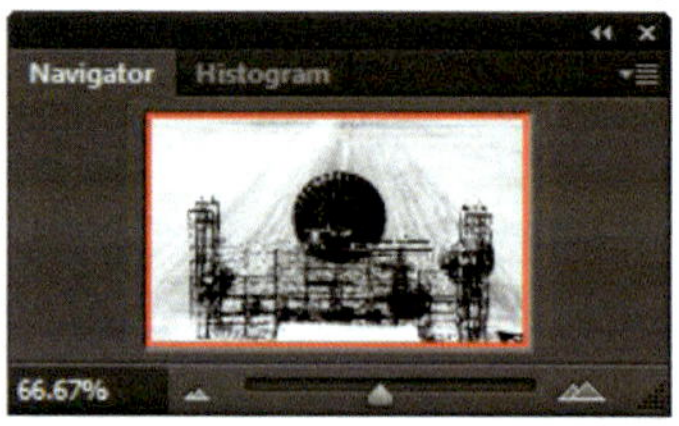

Figure 20.11 - Palette

A palette is functionally similar to a traditional modeless dialog except it does not have to be framed in the dialog chrome and it can be docked to another window. Palettes can be resized, allowing the contents to be reflowed to make best use of available real-estate. By tradition, palettes have a simpler layout than a traditional dialog. Either they contain a small number of controls or the controls are laid out linearly.

Toolbar

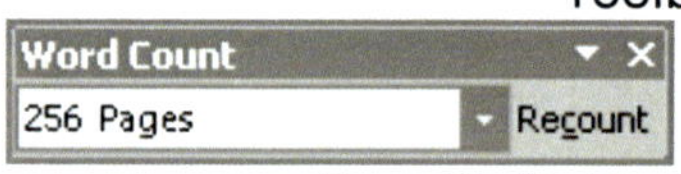

Figure 20.12 - Toolbar

A toolbar is a palette consisting of a collection of related but distinct controls. While a palette may have multiple controls used to represent a single feature, each toolbar control must stand on its own. Toolbars are often customizable, allowing the user to specify the composition and layout of individual controls.

Users have a love/hate relationship with modeless dialogs. On one hand, they like interacting with the controls on their terms rather than when the system demands an answer. They like to pause in the middle of a task to interact with the document surface without having to start the task anew. On the other hand, modeless dialogs can get disconnected from the window that spawned them. While interacting with multiple applications and switching between many windows, it is easy for undocked modeless dialogs to get lost. For example, consider a user bringing up a modeless Find dialog. After a few minutes, the dialog gets hidden behind other windows that got brought up on top. The user then needs to use the Find dialog again so he tries to launch it. Unfortunately it cannot be launched because it is already up, hidden behind another application. To avoid this and similar problems, the following guidelines are to be followed:

- **Be careful of the footprint**. Modeless dialogs are visible longer than their modal cousins. If they float over the document surface, they can cover the user's data. If they are docked, they set aside an entire column or row of real-estate on which the user's data cannot reside. In each case, the risk exists that the data the user cares about will not be visible. To minimize this risk, make the footprint of a modeless dialog as small as possible.
- **Make the purpose obvious**. A modal dialog is only visible shortly after the user spawned it, making it unlikely the user will forget its purpose. A modeless dialog, having much greater longevity, can be "discovered" long

after it was spawned. At all points in time, it must be completely obvious what purpose the dialog fulfills and how its controls can be used to influence the user's scenario.

Document Surface

Most desktop applications have a single window containing the raison d'être (reason or justification of its existence) of the application. This window, called the document surface, is commonly dominated with a representation of the user's data and garnished with controls useful for manipulating the data. The document surface may contain a realistic rendition of the user's data (such as an image manipulation program like Adobe Photoshop or a word processor like Microsoft Word) or an abstract representation (such as a wire-frame or false-color rendition of a car in a Computer Aided Drafting (CAD) application). In other words, there are as many variations of document surfaces as there are applications. Nevertheless, a few guidelines are common to all.

- **Focus on the user's data**. The most prominent screen real-estate, and indeed the vast majority of the screen real-estate, should be devoted to the user's data. Applications failing to adhere to this guideline by devoting too much real-estate to interactive elements are commonly more difficult to use than are necessary.

- **Minimize mouse movements**. An effort should be made to minimize the effort necessary for the user to perform a common task or use-case. Some applications fail miserably at this, forcing the user to cross the screen several times to perform even the simplest actions. Applications with their interactive elements grouped on one side of the screen (typically the top or right side) often do well by this criteria; applications with interactive elements surrounding the document surface typically fail.

- **Make document surface interface items obvious**. Designers are often tempted to add rich interactive features to the document surface that are exposed by right-clicks, clicking on special regions of the screen, or clicks combined with various key combinations. These are notoriously difficult for the user to find. Give the user a chance of finding these elements! Cursor changes, hover effects, and similar clues go a long way to helping the user become an expert.

- **Pay special attention to the user's mental model**. No user will understand the data nor the interface elements as well as you the designer. It is useful to frequently ask yourself "how much must the user know to use this system," and "how much should he know?" Strive always to put the smallest possible demand on the user by presenting the simplest imaginable mental model.

Examples

Example 1

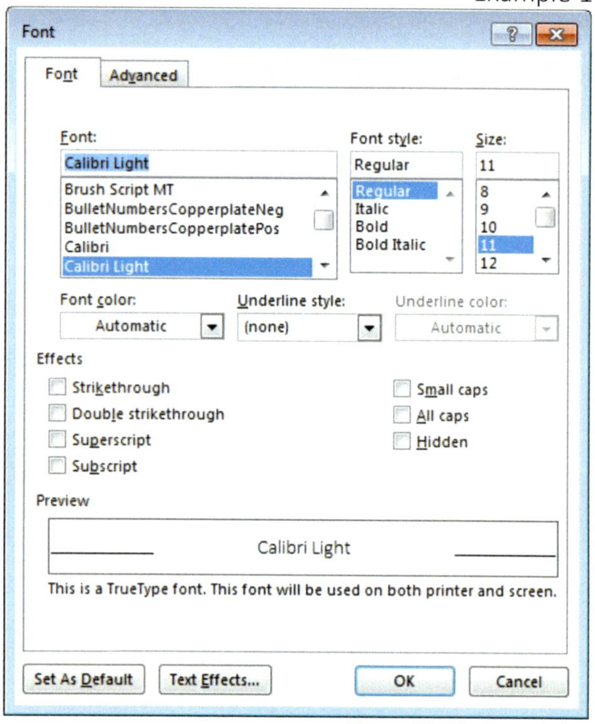

Figure 20.13 - Font dialog

Font Dialog Buttons

Consider the Font dialog in Microsoft Word. Specifically, look at how this dialog uses buttons. Analyze the buttons according to the following desktop standards: Labels and Variants.

Take time to get the labels right

The seven "Effects" checkboxes have noun labels with an implied verb ("Set effect to strikethrough"). There is no confusion as to what the control does nor what the checked state corresponds to.

The [Set As Default] button uses the familiar [Verb Noun] format, being unambiguous as to what it does (performs the action "Set") and to what the effect is applied (default font).

The [Text Effects...] button has an implied verb due to the end ellipse ("Launch the Text Effects dialog"). While it is unclear what text effects are and how they are different than the effects present in this dialog, the user can discover the answer by clicking on the button.

The [OK] and [Cancel] button are standard buttons familiar to all users.

Use the correct variant

The four buttons on the bottom of the dialog have good labels and use the correct variant. One could perhaps that "Set As Default" should be a checkbox.

As for the seven "Effects" checkboxes, a problem exists. Since you can either have "Strikethrough" or "Double strikethrough," these behave like radio buttons. The same is true with "Superscript" and "Subscript" as well as "Small caps" and "All caps." Does this abuse of standards negatively impact usability? Would radio buttons be a better choice? Probably not.

Example 2

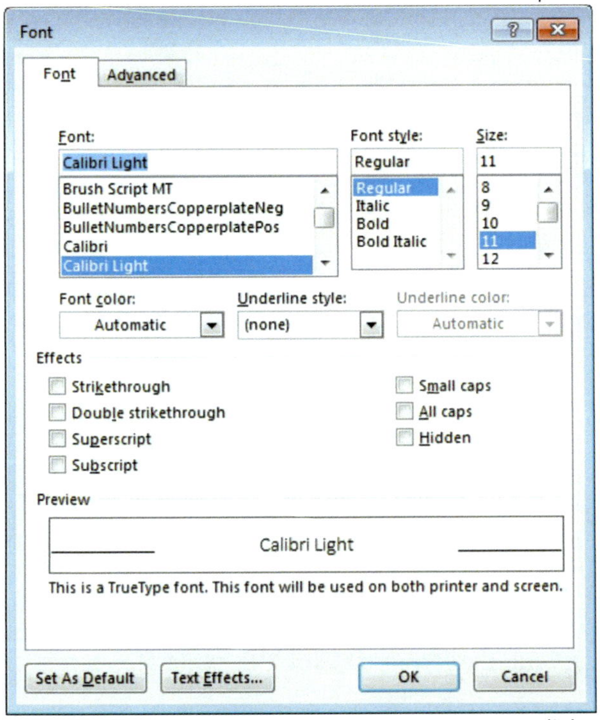

Figure 20.14 - Font dialog

Font Dialog Menu

Consider the Font dialog in Microsoft Word. Specifically, look at how this dialog uses menus. Analyze the buttons according to the following desktop standards: Number of Options, Labels, Defaults, and Variant.

Give the user an appropriate number of selection options

The "Font" control has a ridiculous number of options. Even the most seasoned users feel overwhelmed by the list and cannot internalize all that it offers. Most users search for the handful of fonts they commonly use and ignore the rest.

The "Font style" and "Size" controls have a useable number of options, easy for the average user to internalize.

The "Font color," "Underline style," and "Underline color" controls have a large number of options, but are presented in an easy-to-understand format. No problem here!

Take time to get the label right

All of the labels are clear and concise except one. "Font" is the same label used on the tab and for the overall dialog. Fortunately the contents of the listbox are obvious so a label is not even needed.

Provide defaults whenever possible

None of these controls have defaults in the traditional sense. Instead they reflect the settings of the underlying text.

Use the correct variant

The combo-atomic used in the "Font" and "Size" controls are well chosen; many users know the name of the font or the size they want and simply type it in. The "Font style" control, on the other hand, does not need to be because there are a very small number of settings. Note that some fonts have more than 4 styles and some have none so it cannot be radio buttons.

Example 3

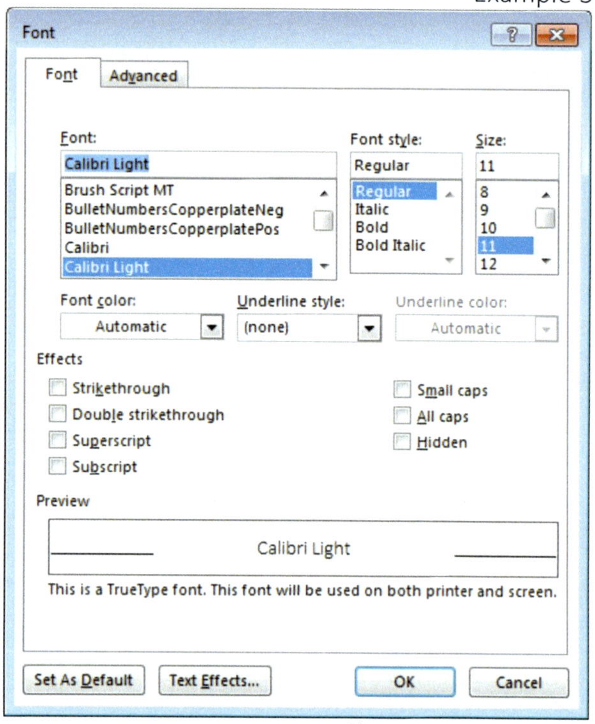

Figure 20.15 - Font dialog

Font Dialog - Modal

Consider the Font dialog in Microsoft Word. Specifically, look at how this dialog uses modality. Analyze it to the following desktop standards: Grouping, Sparingly, Accept/Reject, and Reading Order.

Be intentional about grouping

The two tabs are grouped using the standard tab control, clearly distinguishing what is part of the tab chrome, what is within each tab, and which buttons are common to all tabs (the four buttons on the bottom).

There are three main groups (typeface, effects, and preview). Of these, only the bottom two have explicit grouping controls and labels. There should be a "groupline" (the horizontal line with a label) along the top of the tab.

The controls within the three main groups are also grouped in an intelligent and understandable way. The only possible exception is that the "Effects" group should probably have three sets of controls (Strikethrough and Double strikethrough, Superscript and Subscript, and the remaining three. It turns out that this design was not chosen because there are more controls in Far East and Middle East versions of the dialog.

Use modal dialogs sparingly

All of the functionality available in the Font dialog are available in toolbar buttons or other modeless formats. This makes the font dialog a prime candidate to convert into a modeless dialog.

Make "accept" and "reject" consequences clear

The [OK] button clearly relates to the accept action, meaning the font formatting in the preview will be what is applied to the document surface. No confusion there. The [Cancel] button also clearly relates to the reject consequence.

Reading Order

All of the controls are laid out in a top-down and left-to-right format, following the traditional reading order of a printed page. At the end of the reading order is the [OK] and [Cancel] buttons which is where the user makes a final determination as to whether the formatting will be applied to the document.

Problems

Problem 1 **View Tab - Buttons**

Consider the View tab in Microsoft PowerPoint for the Apple OS X platform. Critique this use of buttons in this example. Pay special attention to the guidelines for button use:

- Take time to get the label right.
- Use the correct variant.

Suggest design alterations that could overcome some of the current design's shortcomings.

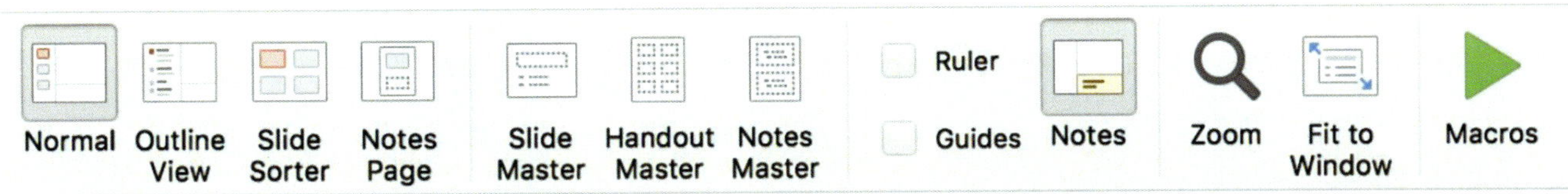

Figure 20.16 - View tab of Microsoft PowerPoint for Apple OS X

Problem 2 **Color Dialog - Menus**

Consider the following color dialog for the Microsoft Windows platform. Critique this use of menus in this example. Pay special attention to the guidelines for menu use:

- Give the user an appropriate number of selection options
- Take time to get the label right
- Provide defaults whenever possible
- Use the correct variant.

Suggest design alterations that could overcome some of the current design's shortcomings.

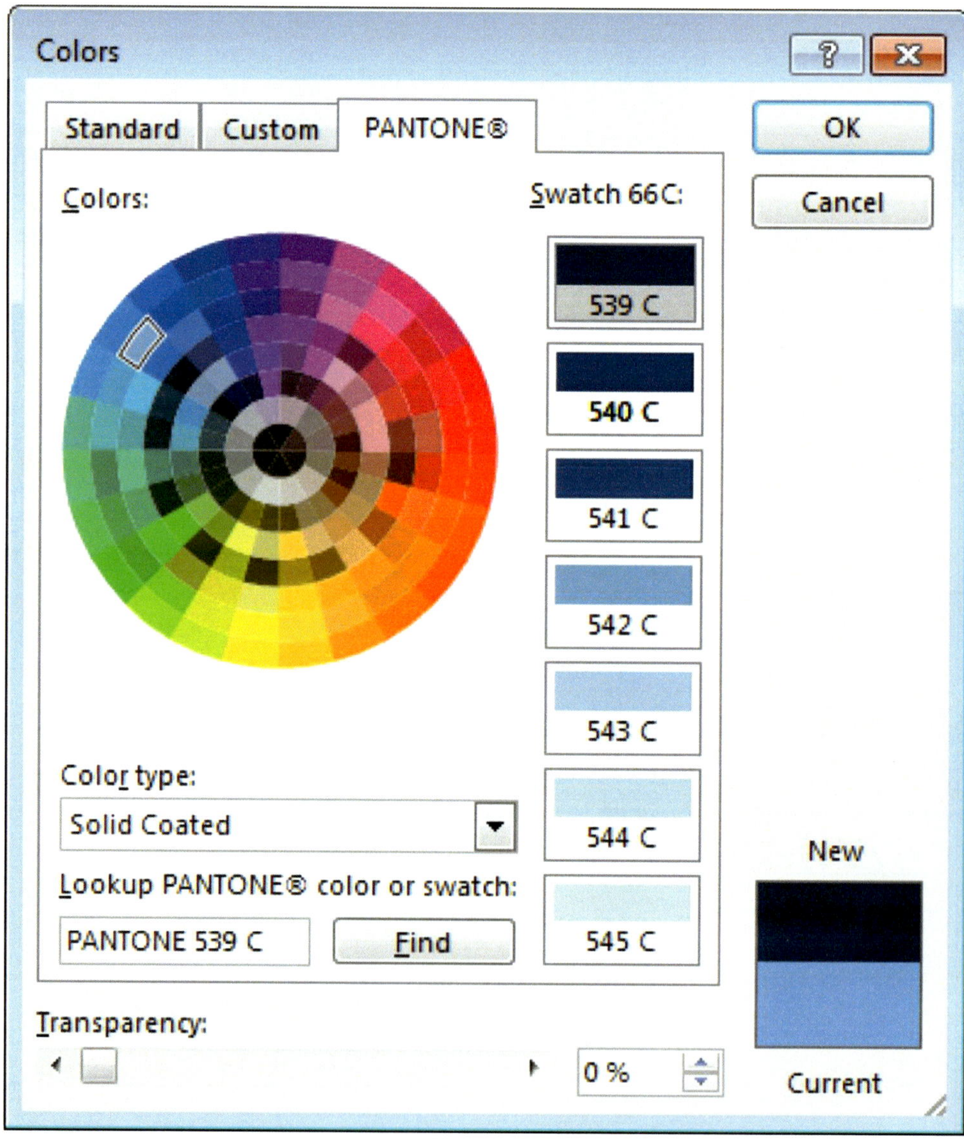

Figure 20.17 -Microsoft Publisher's color dialog

Problem 3 **Create Source Dialog – Edit Controls**

Consider the following dialog for the Microsoft Windows platform. Critique this use of edit controls in this example. Pay special attention to the guidelines for edit control use:

- Make the question clear.
- Make the expectations clear.
- Provide help

Suggest design alterations that could overcome some of the current design's shortcomings.

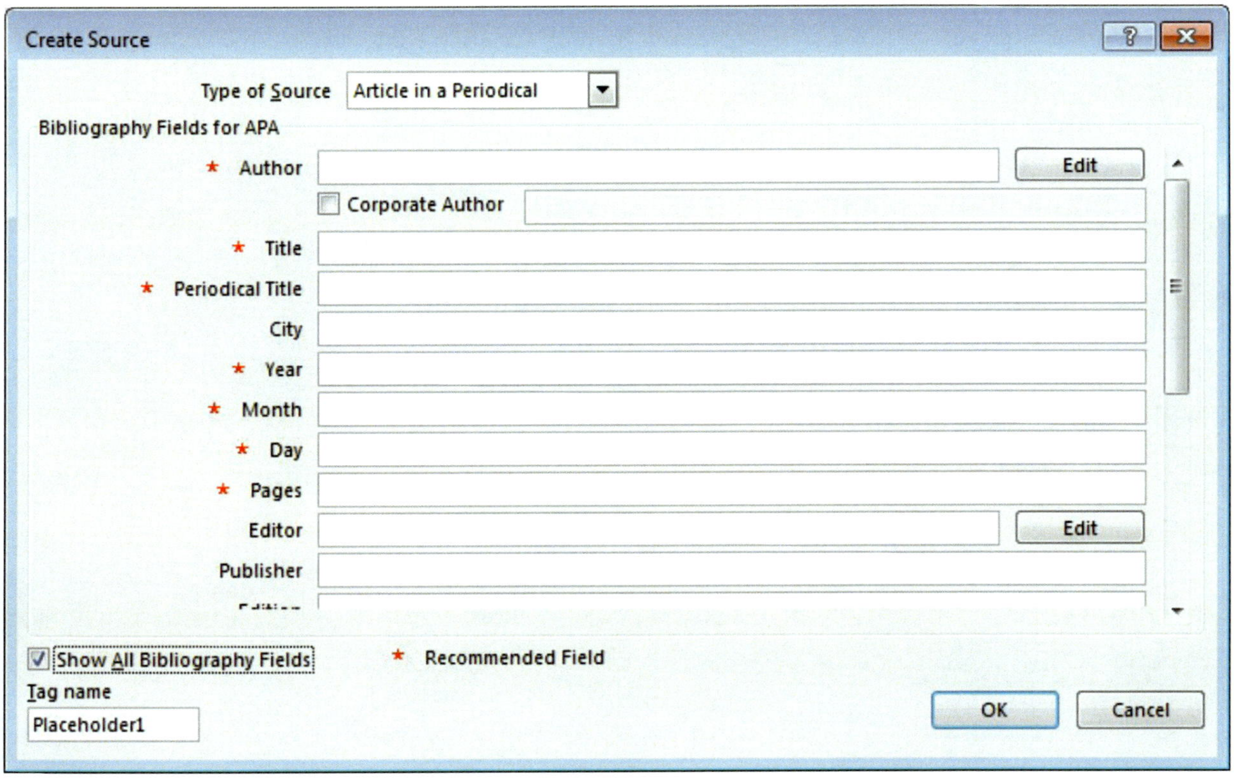

Figure 20.18 - The Create Source dialog as part of the Reference feature in Microsoft Word

Problem 4 **Insert Picture Dialog– Modal Dialog**

Consider the following dialog for the Apple OS X platform used to select a picture to be inserted into a document. Critique this use of modal dialogs in this example. Pay special attention to the guidelines for modal dialog use:

- Be intentional about grouping.
- Use modal dialogs sparingly.
- Make "accept" and "reject" consequences clear.

Suggest design alterations that could overcome some of the current design's shortcomings.

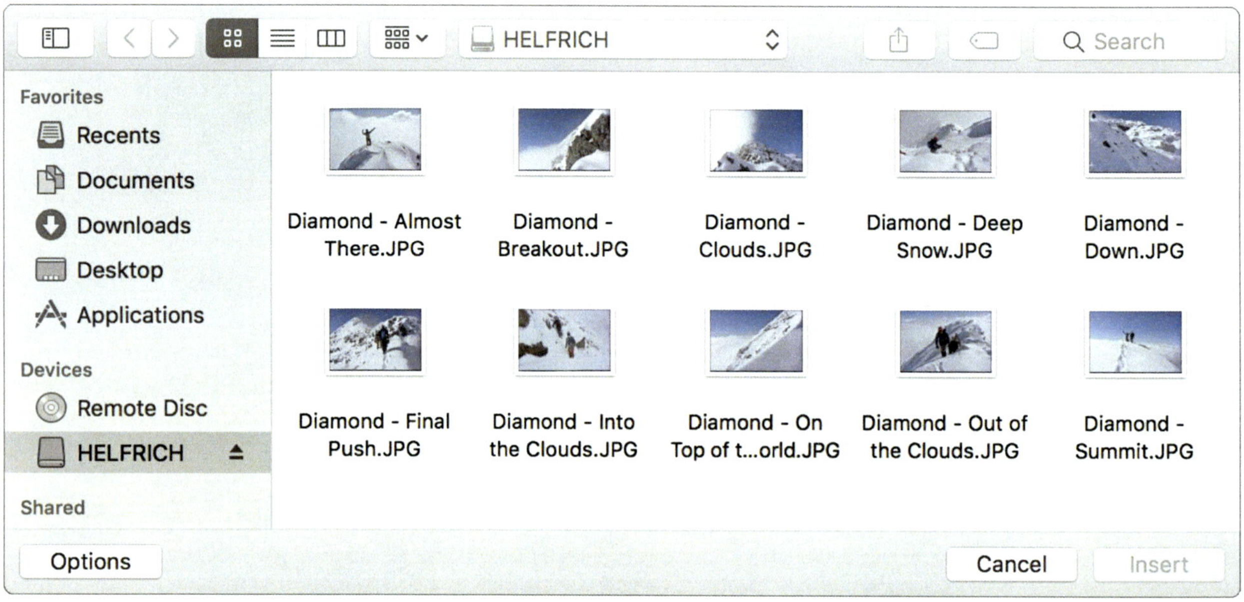

Figure 20.19 - Insert Picture dialog on Apple OS X

Problem 5 **Styles Pane – Modeless Dialog**

Consider the following dialog for the OS X platform. Critique this use of modeless dialogs in this example. Pay special attention to the guidelines for modeless dialog use:

- Be careful of the footprint.
- Make the purpose obvious.

Suggest design alterations that could overcome some of the current design's shortcomings.

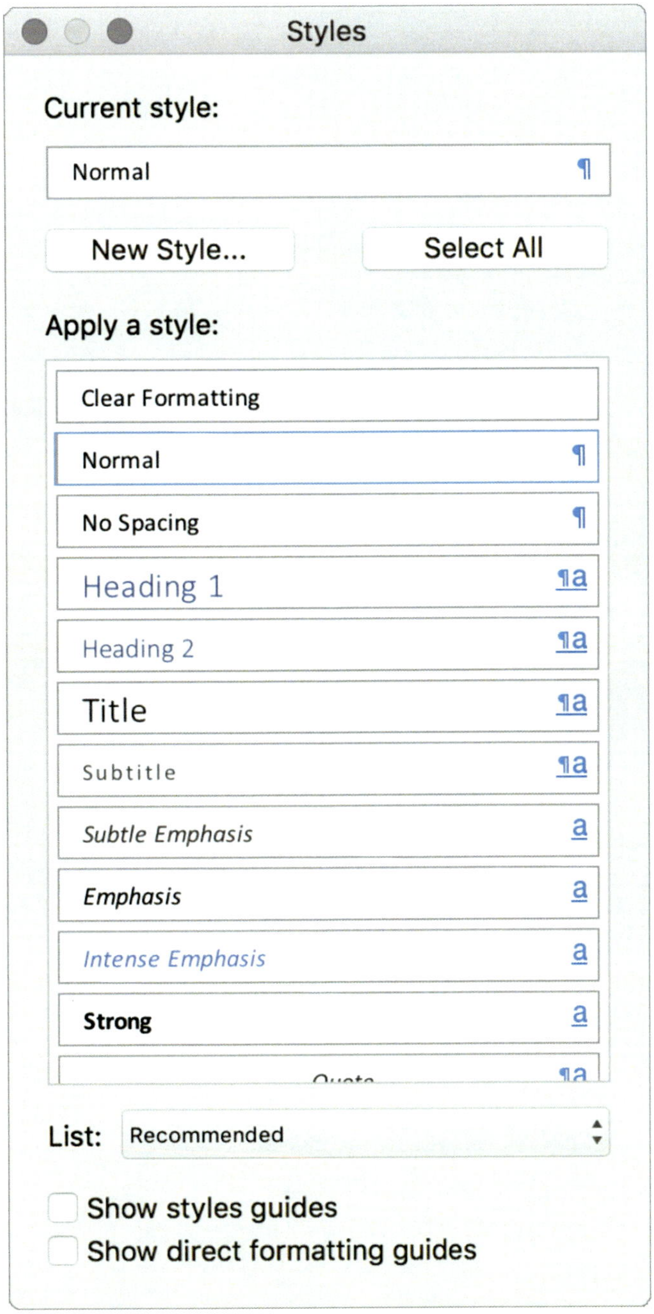

Figure 20.20 - The Styles pane for Microsoft Word on Apple's OS -X

Problem 6 **Word Pad – Document Surface**

Consider the following application for the Windows platform. Critique this use of document surface in this example. Pay special attention to the guidelines for document surface use:

- Focus on the user's data.
- Minimize mouse movements.
- Make document surface interface items obvious.
- Pay special attention to the user's mental model.

Suggest design alterations that could overcome some of the current design's shortcomings.

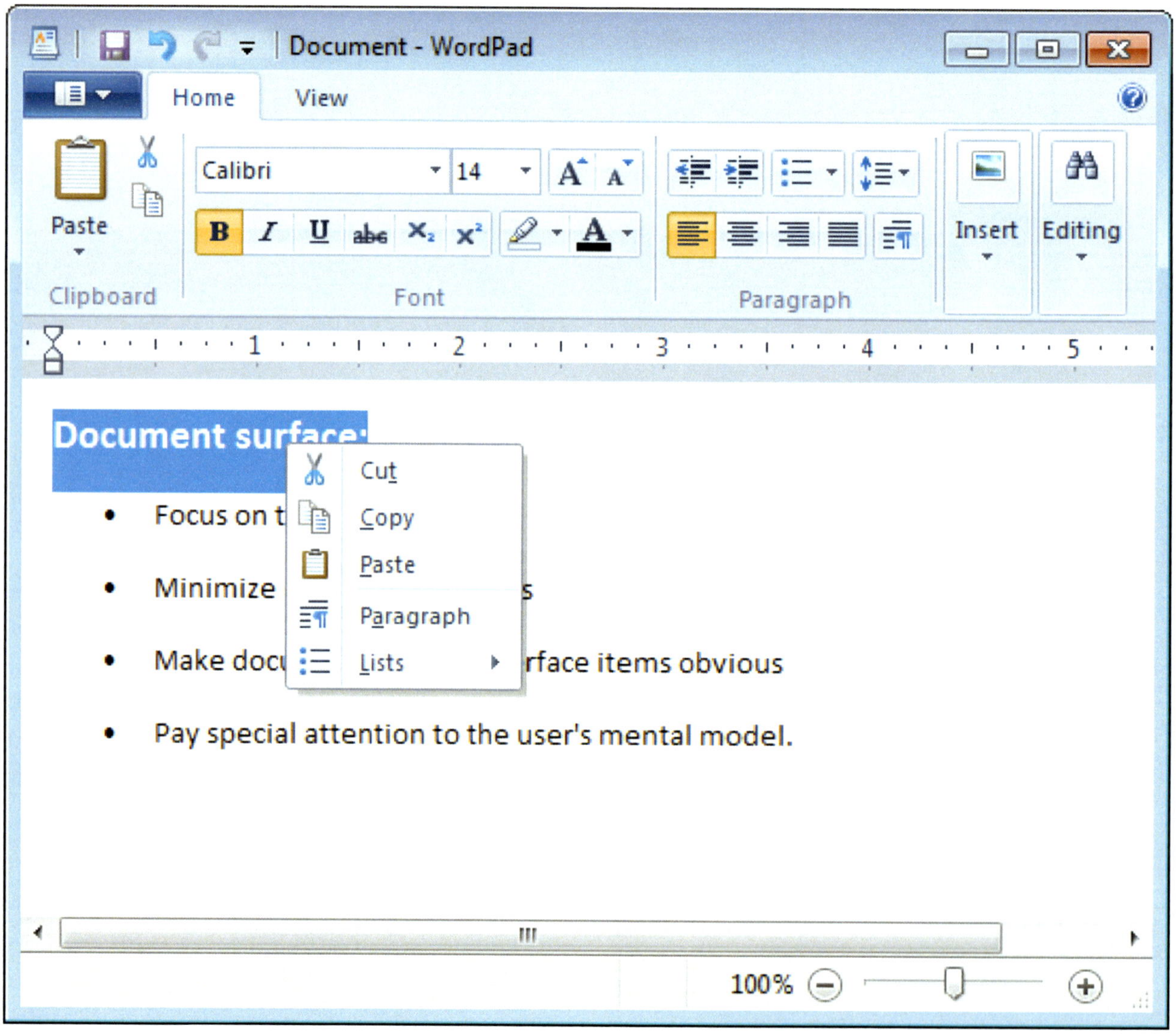

Figure 20.21 - Microsoft WordPad with right-click menu

Appendix

A collection of tools to help us better navigate the topic of HCI and this textbook

Glossary

analogous A color scheme defined as a collection of colors with a small variation in hue while using larger variations of saturation and luminosity

analysis The process of discovering the problems with a given design. This is one of the three parts of the HCI process: analysis → design → verification

balance One of the six principles of layout as they pertain to HCI, balance is the state of a design where it appears even, not appearing to tip over to one side or another

branding Visual clues in a product or user interface designed to inform the user of what is his using and where he is

button A clickable region that launches some action

character spacing The amount of white space between adjacent glyphs

checkbox A flavor of a button exclusively used to reflect state

CMYK A color-space describing color by the amount of Cyan, Magenta, Yellow, and Key (black) colors to add. This is a subtractive color space

coding A qualitative data analysis tool consisting of marks or tags on the transcript of an interview

color space A mechanism for representing color. The most common color spaces are RGB, CMYK, HSL, and C.I.E.

color theory The science of mixing colors so they appear pleasing to the eye

combo-atomic A type of menu control consisting of a standard drop-down listbox with an edit control

compatibility One component of the familiarity variable of usability, compatibility is the measure of how one design resembles another

complementary A color scheme defined as a collection of colors having the same or opposite hue

consistency One component of the familiarity variable of usability, consistency is the measure of how one part of a design resembles another

content page The content pages of a web site contain all the data and functionality of the site. The purpose of home pages and menu pages are to direct users to the content pages they care about

contrast One of the six principles of layout as they pertain to HCI, contrast is the process of creating a pattern and then breaking it. The eye is naturally drawn to the item that does not fit the pattern

coversheet A qualitative data analysis tool consisting of a single sheet of paper summarizing all that was learned from a given participant or interview

criteria Another name for "evaluation criteria"

customer An individual who makes the purchase decision for a product. This is not the same thing as the person who uses the product (user)

deep dive A process of taking a design all the way to the end of the refinement process to see how it plays out. This is done to get a feel for the fitness landscape

design The process of channeling creative processes to maximize the chance a good design will be found. This is one of the three parts of the HCI process: analysis → design → verification

diphthong A diphthong is a combination of letters used to symbolize a linked sound

directness One of the measures of the efficiency of a design, it is the ability of a system to accept input from the user to perform a specific action with the least amount of effort

discoverability One component of the visibility variable of usability, discoverability is the probability that a given user will be able to find a given piece of information

edit control A control allowing the user to provide free-form or loosely-constrained input through the keyboard

efficiency The amount of effort or time required to perform a task. This is one of the eight variables of usability. Additionally, efficiency is one of the three measurements of the quality of an assessment instrument. The other two are validity and reliability

elevator pitch Relating the essence of a feature to someone in thirty seconds with no visual aids. In other words, describing something while standing in an elevator

enclosing A grouping tool where related items are enclosed in a box, bounded by a line, or set apart by another visual element

ergonomics The process of adapting the work environment to fit the needs of the human frame. This is more Human Factors than HCI

evaluation The process of the user interpreting the output from the interface. This is half of the interaction framework: evaluation & execution. It is also the name for the fourth part of the evaluation side of the interaction framework: presentation → perception → interpretation → evaluation. In that context evaluation refers to the process of the user applying knowledge gained from the user interface to their overall understanding of the system

evaluation criteria A collection of constraints in the design process through which various design alternatives are measured. All criteria are expressed in terms of the users and can be traced to a variable of usability

execute sequence The user knows what needs to be done (specify sequence) and physically attempts to perform the task. This is the fourth step of the execution side of the interaction framework: goal → intention → specify sequence → execute sequence

execution The process of the user expressing his needs to the input channels afforded by the interface. This is half of the interaction framework: evaluation & execution

executive summary A short description of what a feature or product is all about. This is usually the first section of a specification

experimental analysis One of the three evaluation techniques, Experimental Analysis consists of evaluating the fruits of the labors

expert analysis One of the three evaluation techniques, expert analysis consists of known expert giving the design a once-over

familiarity One of the eight variables of usability, familiarity is the degree in which the interface resembles something the user has used before

fault tolerance One component of the trust variable of usability, fault tolerance is how resistant the system is to failures introduced by the user or external sources

feature A description of one aspect of a project

flexibility One of the measures of efficiency for a design, it is the ability of a system to accept input from the user on his terms.

fitness landscape A multi-dimensional space where each point is characterized with a value representing the quality or fitness of the design at that point. The fitness landscape is a tool to help us with the design process by helping us visualize the set of all possible design alternatives

Fitt's law The closer the control and the larger the control the easier it will be for the user to select it. This works both with Human Factors and HCI

futz A form of escapism where the user unintentionally dwells on unimportant details rather than tackling difficult or unpleasant tasks

goal What the user is trying to accomplish. This is the first step of the execution side of the interaction framework: goal → intention → specify sequence → execute sequence

glyph A glyph is a single manifestation of a letter, comprised of one or more strokes

grouping One of the six principles of layout as they pertain to HCI, grouping is the process of encouraging the user to infer a relationship exists between design elements

gulf of evaluation The failure of the user to accurately perceive or interpret the state of the system. This, coupled with the gulf of execution, constitute the two sources of errors in the interaction framework

gulf of execution The inability of the user to do what he wishes to do with the system. This, coupled with the gulf of evaluation, constitute the two sources of errors in the interaction framework

heuristic analysis One of the three evaluation techniques, heuristic analysis consists of studying the pieces of the design

hill climbing A design process resulting from the uniform slope property of the fitness landscape where changes in a design continue in a given direction until the quality of the design no longer improves

hit region A flavor of a button that is characterized by a clickable region of an image

home page The top-level page of a web site serving to direct users to the content they care about, to heighten their interest in the site, and to advertise the purpose of the entire web site

HSL A color space where each color is defined by the Hue, Saturation, and Luminosity necessary to describe a given color

hyperlink A flavor of a button that is characterized with a textual label that is laid out the same as inert text

intention What the user is trying to accomplish taking the system into account. This is the second step of the execution side of the interaction framework: goal → intention → specify sequence → execute sequence

interactive mockup A type of prototype built with technology like Microsoft PowerPoint, Adobe Flash, or HTML

interpretation The process of the user understanding the symbols and metaphors used by the system. This is the third part of the evaluation side of the interaction framework: presentation → perception → interpretation → evaluation

kerning The process of adjusting the spacing of adjacent glyphs so their spacing appears consistent and uniform

LAB A color-space designed to represent how people perceive color.

labeled button A flavor of a button that is characterized by a bounding rectangle and an image or text label describing the button's purpose

layman Unskilled worker or individual without specific training

learnability One of the eight variables of usability, learnability is the path for a novice to become an expert

letter A letter is a symbol corresponding to a spoken sound. A letter has semantic meaning

ligature A ligature is the combination of two or more overlapping glyphs into a single glyph done for the purpose of increasing readability

listbox A type of menu control characterized by a set of options usually represented with noun text labels

mapping One of the eight variables of usability, mapping is the process of how the interface communicates the system mental model to the user so he or she can form a valid mental model

mental model A user's understanding of how a system works. A mental model includes the individual components of a system, an understanding of how the components work, and the relationship between components. The size of a mental model is measured by the simplicity variable of usability. The degree in which an interface communicates a valid mental model is the mapping variable of usability

menu A list of related buttons designed to facilitate the user selecting an option from a pre-specified list

menu page A menu page serves the same purpose as a home page except on a smaller scale: the menu page represents a sub-set of the larger web site

modal dialog A window associated with an application designed to help the user perform some task. Modal dialogs take focus from the underlying window forcing the user to complete the task before working with the rest of the application

modeless dialog A window associated with an application designed to help the user perform some task while allowing interaction with the rest of the application

monochromatic A color scheme are collections of colors with a constant hue but different degrees of saturation and luminosity

motivation One of the eight variables of usability, motivation is the degree in which the user wants to use the system

multi-select listbox A type of menu control. The multi-select listbox is the same as a standard listbox except multiple items can be selected

numeric input A type of edit control optimized to allow for selection of scalar values. This is also known as a spinner

palette A type of modeless dialog providing a rectangle of real-estate on which user interface elements may be placed. Palettes typically can float over the underlying document surface or be docked to the side of the document surface

paper prototype A type of prototype consisting of a mockup of an interface complete with semi-interactive elements created out of paper or a similar convenient medium

partial working system A type of prototype built with the same technology used to make the final product

perception The physical ability of the user to be able to accept the input the interface is presenting. This is the second part of the evaluation side of the interaction framework: presentation → perception → interpretation → evaluation

persona A hypothetical archetype representing either the entire target audience or a significant portion of the target audience

precedence One component of the visibility variable of usability, precedence is a measure of how prominent a given item is in the user interface

presentation The system correctly reflecting the user's data or the state of the system through some interface elements. This is the first part of the evaluation side of the interaction framework: presentation → perception → interpretation → evaluation

pre-structured case outline A qualitative data analysis technique consisting of a representation of the researcher's conceptual framework about the data that will be collected

proximity A grouping tool where a relationship between items is implied by their relative position to each other. The closer they are to each other, the greater the implied relationship

qualitative data Data in the form of words or textual descriptions

quantitative data Data in the form of numbers that can be counted

radio group A type of menu control, a radio-group is a list of checkboxes where exactly one value can be selected at a time

reachability One component of the visibility variable of usability, reachability is how accessible a given piece of information or functionality is to the user

reading order One of the six principles of layout as they pertain to HCI, reading order is the path the user's eyes follow as it moves through a design

reliability One of the three qualities of an assessment instrument, reliability is the amount of error in a measurement. In other words, if a second measurement is taken of the same participant, will the resulting data be the same as the first?

requirement A requirement is a list of what must be present in a project. All requirements are specified in terms of the project or the technology, not in terms of what the user is trying to do

responsiveness One tool to increase efficiency of a design, responsiveness is how quickly the design responds to user action

RGB The color space designating color according to the amount of Red, Green, and Blue light mixed to describe a given color. This is an additive color space

rich text A type of edit control designed to allow the user to enter more than one line of input and to allow for character (and often paragraph level) formatting

rhythm One of the six principles of layout as they pertain to HCI, rhythm is a strong, regular, and repeated pattern

rubric A quantification tool, enabling us to derive a score from qualitative data according to pre-established criteria

scenario Scenario is a story representing an important use-case of the user interacting with the system

scientific management The process of increasing work efficiency through analysis of workflow

serif A small line or embellishment at the end of a letterform designed to increase readability

shade A variation of a color defined by adding black. This serves to decrease luminosity and decrease saturation

simplicity One of the eight variables of usability, simplicity is the size and complexity of the user's mental model of the system

slant The angle of stress in a glyph

specification A document serving as the primary communication vehicle between members of the design team

specify sequence The process of the user formulating a plan to carry out his intention. This is the third step of the execution side of the interaction framework: goal → intention → specify sequence → execute sequence

split complementary A collection of colors defined as a mixture of a complementary and analogous color scheme

stakeholder An individual with a vested interest in a design. This could be a user, a customer, or even the people building a product

stress The ratio between the thinnest and thickest part of a bowl on a letterform

storyboard A type of prototype consisting of a collection of drawn screens representing a single scenario

target audience The group of people for whom a design is intended. In other words, a collection of users

task saturation The condition when the user does not have the time to perform the necessary tasks even when he knows how

template A set of guidelines giving the content page designer freedom to explore variations in the layout of a page

tetradic A color scheme defined as a combination of two complementary color collections

textbox A type of edit control used to allow the user to input a single line of text

tint A variation of color defined by adding white. This serves to increase luminosity and decrease saturation

tone A variation of color defined by decreasing saturation while keeping the luminosity and hue constant

toolbar A toolbar is a modeless dialog consisting of a collection of related but distinct controls

toolbar button A flavor of a button typically drawn with an icon instead of text. Toolbar buttons are frequently used to reflect state

triadic A collection of colors separated by 120 degrees on the color wheel

trust One of the eight variables of usability, trust is how confident the use feels when interacting with the system

unity One of the six principles of layout as they pertain to HCI, unity is the process of making all the items in a design appear to belong to a greater whole

use-case Related to a scenario, a use-case is one instance of a user carrying out his Intention with the system

user An individual who uses a product. Not to be confused with the individual who makes the purchase decision (customer)

validity One of the three qualities of an assessment instrument, validity is about measuring the right thing.

value proposition The rational why a customer or user will be compelled to use a product. This is typically an important part of an executive summary

variables of usability A collection of variables constituting the heuristics by which an interface design can be criticized. These variables are: efficiency, learnability, familiarity, simplicity, mapping, motivation, trust, and visibility

verification The process of validating all the assumptions made during the analysis and design process

vignette A qualitative data analysis tool consisting of a focused description of a series of events taken to be representative, typical, or emblematic

visibility One of the eight variables of usability, visibility is a measure of how well the system presents all the information and functionality of the system to the user when it is needed

visual similarity A grouping tool where a relationship between items is implied because they have a similar size, shape, coloring, or drawing style

white space Otherwise known as negative space, white space is one method to achieve contrast in a design. It is the process of drawing attention to a part of a layout where less controls or design elements exist

Wizard of Oz A type of prototype where the user is presented with a rendition of the interface and "clicks" on interactive elements with her finger. An unseen manipulator then alters the rendition to give the perception of responsiveness

zero-sum-game A truism stemming from the visibility variable of usability: the more precedence you give some data or feature, the more you are taking away from another

Works Cited

Anthony T. (2012, February 21). *Finger-friendly design: Ideal mobile touchscreen target sizes.* Retrieved from Smashing Magazine: http://uxdesign.smashingmagazine.com/2012/02/21/finger-friendly-design-ideal-mobile-touchscreen-target-sizes/

Apple Inc. (2010). *Human Interface Design Principles.* Retrieved from Apple Human Interface Guidelines: http://developer.apple.com/mac/library/documentation/UserExperience/Conceptual/AppleHIGuidelines/XHIGIntro/XHIGIntro.html

Apple Inc. (2012, 7 23). *UI Element Guidelines: Controls*. Retrieved from OS X Human Interface Guidelines: https://developer.apple.com/library/mac/#documentation/UserExperience/Conceptual/AppleHIGuidelines/Controls/Controls.html#//apple_ref/doc/uid/TP30000359-TPXREF141

Asimov, I. (1942, March). Runaround. *Astounding Science Fiction* .

Barker, P. (1994). Designing interactive learning. In T. de Jong, & L. Sarti, *Design and Production of Multimedia and Simulation-based Learning Material* (pp. 1-30). Dordrecht, The Netherlands: Kluwer Academic Publishers. Retrieved from: http://books.google.com/books?hl=en&lr=&id=AQUN0StHHPIC&oi=fnd&pg=PA1&dq=Barker+%22designing+interactive+learning%22&ots=momz2Zvd2e&sig=W2FF63rWWHDIAiGoKywL6OldiPc#v=onepage&q=Barker%20%22designing%20interactive.

Bringhurst, R. (2002). *The Elements of Typographic Style.* Vancouver, BC: Hartley & Marks, Publishers.

Clark, R. (2008). *Writing Tools: 50 Essential Strategies for Every Writer.* Little, Brown, and Company.

Cooper, A. (2004). *The Inmates are Running the Asylum.* Indianapolis: Sams Publishing.

Dandekar, K., Raju, B., & Srinivasan, M. (2003). 3-D finite-element models of human and monkey fingertips to investigate the mechanics of tactile sense. *Transactions of the ASME*, 682-691. doi:10.1115/1.1613673

Dvorak, A. (1936). *Typewriting Behavior.* American Book Company.

Fitts, P. (1954). The information capacity of the human motor system in controlling the amplitude of movement. *Journal of Experimental Psychology, 47*(6), 381-391.

Fitts, P., & Jones, R. (1947). Analysis of factors contributing to 460 "pilot error" experiences in operating aircraft controls (Report No. TSEAA-694-12). *Air Materiel Command, U.S. Air Force*.

Ford, A., & Roberts, A. (1998, 11 1). *Colour Space Conversions.* Retrieved from http://www.poynton.com/PDFs/coloureq.pdf

Gilbreth, F., & Gilbreth, L. (1917). *Applied motion study: A collection of papers on the efficient method of industrial preparedness.* New York, N.Y.: Sturgis & Walton Company.

Head, A. (2003, August). *Personas: Setting the stage for building usable information sites.* Retrieved from Information Today: http://www.infotoday.com/online/jul03/head.shtml

Heim, S. (2007). *The Resonant Interface.* San Francisco: Addison Wesley.

Helfrich, J. (2013). Variables of Usability. *HCI International 2013* (pp. 114-118). Las Vegas: Springer.

Hudson, W. (2013). Card Sorting. In M. Soegaard, *The Encyclopedia of Human-Computer Interaction, 2nd Ed.* Retrieved from http://www.interaction-design.org/encyclopedia/card_sorting.html

Leventhal, L. M., & Barnes, J. A. (2008). *Usability Engineering: Process, Products, and Examples.* Upper Saddle River, NJ: Pearson Prentice Hall.

Lynch, P., & Horton, S. (2009). *Interface design*. Retrieved from Web Style Guide, 3rd Edition: http://webstyleguide.com/wsg3/4-interface-design/index.html

Mayhew, D. J. (1992). *Principles and Guidelines in Software User Interface Design.* Upper Saddle River, NJ: Prentice-Hall, Inc.

Microsoft Corp. (2007). *How to design a great user experience.* Retrieved 2007, from Microsoft Developers Network: http://msdn.microsoft.com/en-us/library/aa511335.aspx

Microsoft Corp. (2013, 5 14). *Command Buttons*. Retrieved from Dev Center - Desktop: http://msdn.microsoft.com/en-us/library/windows/desktop/aa511453.aspx

Miles, M., & Huberman, A. (1994). *Qualitative Data Analysis, Second edition.* Thousand Oaks, CA: SAGE Publishing, Inc.

Miller, D. (1981). The depth/breadth tradeoff in hierarchical computer menus. *Proceedings of the Human Factors Society 25th Annual Meeting*, 296-300.

Miller, G. (1955). The magical number seven, plus or minus two: Some limits on our capacity for processing information. *Psychological Review*, 343-352.

Mollerup, P. (2006). Simplicity. *Design Research Quarterly*, 7-15. Retrieved from: http://hdl.handle.net/1959.3/71506.

Norman, D. (1999). Affordance, conventions, and design. *Interactions, 6*(3), 38-43. ISSN: 1072-5520.

Norman, D. (2007). *Simplicity is highly overrated.* Retrieved from The Nielsen Norman Group: http://www.jnd.org/dn.mss/simplicity_is_highly.html

Norman, D. A. (1988). *The Psychology of Everyday Things.* New York: Basic Books, Inc.

Raskin, J. (2000). *The Humane Interface.* New York: Addison-Wesley.

Spolsky, J. (2000, October 15). *Painless Functional Specification*. Retrieved from Joel on Software: http://www.joelonsoftware.com/articles/fog0000000033.html

Taylor, F. W. (1903). Shop management. *American Society of Mechanical Engineers*, . Retrieved from: http://www.gutenberg.org/etext/6464.

Worthen, B., White, K., Fan, X., & Sudweeks, R. (1999). *Measurement and Assessment in Schools.* New York, NY: Addison Wesley Longman, Inc.

Index

analogous 190
balance 175
branding 220
button 245
checkbox 245
CMYK 188
coding 168
color space 188
color theory 187
combo atomic 246
complementary 191
contrast 174
coversheet 166
customer 43, 49, 50
deep dive 102
dropdown listbox 246
edit control 247
efficiency 25, 122, 221, 231
enclosing 172
ergonomics 7
evaluation 15, 17
evaluation criteria 68, 99, 103, 123, 125, 127, 128, 137, 142
execute sequence 20
execution 15
executive summary 90, 103, 126, 128
experimental analysis 79
expert analysis 79
familiarity 28, 123, 231
fault tolerance 33
feature 67
fitness landscape 99, 102
Fitt's law 6
futz 8
goal 18
grouping 172
gulf of evaluation 15, 47, 60
gulf of execution 18, 47, 60
heuristic analysis 79
heuristic evaluation 142
hill climbing 99, 100
hit region 245
home page 221
HSL 189
hue 190
hyperlink 245
intention 19, 223
interaction framework 60
interactive mockup 110, 119
interpretation 17
interviewing 45
learnability 26, 122
listbox 246
luminosity 190
mapping 30, 123, 172, 174, 221, 222, 223
mental model 29
menu 246
menu page 222
modal dialog 248
modeless dialog 249
monochromatic 190
motivation 31, 123, 221, 222, 231
multi-select listbox 246
numeric input 247
palette 249
paper prototype 111, 118
perception 16
persona 59, 69, 123, 124, 125, 126, 128, 141, 158, 164, 169
presentation 16
pre-structured case outline 165
prototype 127
proximity 173
radio group 246
reachability 34
reading order 177
reliability 143
requirement 67
RGB 188
rhythm 178, 223
rich text 247
rubric 71
saturation 190
scenario 57, 69, 110, 115, 119, 123, 125, 127, 128, 141, 158
scientific management 5
shade 190
simplicity 29, 123
specify sequence 19, 59
split complementary 191
stakeholder 43, 49

storyboard 110, 117
target audience 43
task saturation 6
tetradic 191
text box 247
tint 190
tone 190
toolbar 249
toolbar button 245
trust 32, 123
unity 176, 223
use-case 56, 112, 246, 248, 250
user 43, 49, 50
validity 143
value proposition 91, 94
variables of usability 25, 47, 69, 82, 85, 86, 122, 165
vignette 169
visibility 34, 123, 221, 222, 223, 231
visual similarity 173
white space 174
zero-sum-game 34